'A most impressive work. Travellers to Iceland in the nineteenth century did not see the land as it was, but more as they expected it to be: an exotic landscape. Chris Caseldine takes a different approach, travelling through the history and topography of Iceland with great understanding and knowledge of his subject. It is highly recommended to anyone who loves to explore the unknown.'

– Sigurður Gylfi Magnússon, Professor of Cultural History, University of Iceland

'Anyone whose dreams are haunted by Iceland's mountains and mossy lava plains will find lots to learn and to enjoy in this colorful book and the striking tales it tells. Put on your coziest wool sweater and learn how Iceland's landscapes inspired great debates in earth science and shaped the fates of Icelanders and countless awestruck visitors through the centuries.'

– Yarrow Axford, polar geoscientist and Associate Professor in the Department of Earth and Planetary Sciences, Northwestern University

D0944644

MOST UNIMAGINABLY
STRANGE

MOST UNIMAGINABLY STRANGE

AN ECLECTIC COMPANION
TO THE LANDSCAPE OF ICELAND

Chris Caseldine

REAKTION BOOKS

FOR HELEN, THOMAS, ANNA, LUCY AND SOPHIE

Published by
REAKTION BOOKS LTD
Unit 32, Waterside
44–48 Wharf Road
London N1 7UX, UK
www.reaktionbooks.co.uk

First published 2021

Printed and bound in India by Replika Press Pvt. Ltd

A catalogue record for this book is available from the British Library

ISBN 978 1 78914 472 7

Cover image: Aerial view of Lakagígar, Iceland, source fissure
of the 1783–4 eruption. Photo MartinM303/iStock.com.

Endpapers: Iceland's Fagradalsfjall Volcano Eruption 2021.
Photo Svala Louise Vera Leaman

INTRODUCTION

On Saturday 22 July 1871 the artist and writer William Morris, described by one editor of his work as an eager and polite guest in Iceland, was travelling on horseback across Markarfljót towards Þórsmörk and found himself confronted by a landscape he described as 'most unimaginably strange'. Although yet to see Geysir and Þingvellir, Morris had already ridden from Reykjavík across southern Iceland but was taken aback by finding himself in a landscape so diverse, chaotic and intimidating that he went to great lengths in his diary to try to give an impression of something which his readers would clearly have been unable to grasp, coming as they did from very different landscapes in Britain and Europe. His first sight of eastern and southern Iceland sailing from the Faroes challenged his literary abilities: 'it is no use trying to describe it, but it was quite up to my utmost expectations as to strangeness.' This combination of novelty, awe, apparent chaos, fear, curiosity and uncertainty lies at the heart of the fascination the Icelandic landscape has exerted on travellers, scientists, artists and writers over the centuries.

Coming to terms with the Icelandic landscape, and especially trying to understand why it looks like it does, its geological origins and present form, may require a good understanding of the broad range of earth sciences, but also demands an ability to use language and observations beyond the strictly scientific. In a commentary on his approach to translating the recently published *Öræfi*, by Ófeigur Sigurðsson, Lytton Smith observes that 'we see the humans who confront its geology, topography and geography attempting to find the language that would contain its wonder, tumult and extremes,' supporting the idea that science alone cannot provide an understanding of what Iceland presents.[2] Within accounts by British visitors a recurrent theme is trying to see Iceland as a more intense and exaggerated Scotland, but the added element of strangeness defeats such a straightforward comparison. The challenge for this book is to provide an overview of the Icelandic landscape that goes beyond a simple geological explanation, reflecting what travellers and literary visitors have observed and showing how Iceland has often played an important role in earth science debates of the day. Scientists and travellers who have visited and written about Iceland, as well as Icelandic scientists and non-scientists who rarely got the international recognition they deserved, provide the underpinning for the following pages. While aiming to present a current and scientifically informed view of why Iceland looks the way it does, it is hoped that a sense of wonder and fascination is not lost along the way. Iceland is a challenging country, 'Europe's last wilderness', and not just because of the weather. In terms of location Iceland may be an 'outgrowth of an already insignificant headland', but the country and its landscape has an importance and fascination that far outstrips its size and location.[3] The treatment here is largely, but not exclusively, a view from the outside, and is grounded in the earth sciences, but it is also eclectic as a range of diverse and highly individualistic characters appear, drawn by the unique landscape and culture to be found in the middle of the North Atlantic.

In the earliest accounts and records of those visiting a country considered to be at the very edge of civilization, at least from a European perspective, a number of terms recur. Adam of Bremen, in the eleventh century, describes people living in 'holy simplicity' on an island where ice makes the sea solid and there is a curious form of black ice that is so dry it will burn. As early as 1200,[4] Saxo Grammaticus, the Danish historian and theologian, in his introduction to *Gesta Danorum*, similarly records the Icelanders' simplicity as well as their literacy and in recounting the many marvels of the land expresses surprise that a place so cold could have so many eternally burning fires with an inexhaustible source of fuel. Despite the idea of a land full of marvels and simple, learned folk, writers also emphasized the more challenging aspects of these northern lands. The German Sebastian Franck, in the early sixteenth century, emphasized the constant presence of snow and ice, entire mountains made of ice and a landscape dominated by barren deserts – a view reinforced by the Scot Patrick Gordon a century later in his widely read *Geography Anatomiz'd*: a country 'incumbered with Deserts, barren mountains, or formidable rocks'.[5] In the eighteenth and nineteenth centuries words such as 'awful', 'desolate', 'terrible', 'fearsome' and 'barren' frequently recur in accounts from travellers. There is also a persistent otherworldliness and geographical novelty attached to the country. The Moravian priest Daniel Vetter (also known as Daniel Strejc, 1592–1669) visited the island in 1613 and is

Markarfljót, southern Iceland, looking south towards the Atlantic Ocean; the location for William Morris's comments on a landscape 'most unimaginably strange'.

considered one of the more reliable earlier chroniclers. He included in the title of his account *Islandia*, published in Polish in 1638, the words variously translated as: 'the island of Iceland, where are found things curious (wild) and peculiar, without precedent in our own country'.[6] The redoubtable Austrian traveller Ida Pfeiffer, 'the female Humboldt', who spent three months in the country in 1845, felt 'impelled' to visit the region and 'hoped to see things which should fill me with new and inexpressible astonishment'. In this she was not disappointed. When approaching Reykholt she observed that 'the eye wanders over the vast desert, and finds not one familiar object on which it can rest', and near Mount Hekla she describes the landscape as particularly Icelandic, 'so strange and remarkable, that it will forever remain impressed on my memory'.[7]

Visitors spent many pages recounting the way of life of these people living at the very margins of possible existence (the Danish authorities did after all consider the abandonment of their colony following the Laki eruption in 1783–4), usually concentrating on their diet and a perceived lack of manners;[8] there is also often a feeling of surprise that they were so well read and informed about wider scientific and political matters. Few travelling to Iceland spoke Icelandic or Danish but found themselves able to converse with clerics in Latin, as most were well educated, usually in Denmark. This astonishment, not solely restricted to contacts with priests, is well summarized by John Barrow in 1834:

> We have it on the authority of former travellers in this country, and it is confirmed by the resident Danish merchants of Iceland, that it is no uncommon thing to meet with men labouring in the fields, mowing the hay, digging turf, building the walls of their cottages, sheds, cow-houses, and performing every menial labour, who will write Latin, not merely with grammatical accuracy, but even with elegance.[9]

Not all early travellers' accounts are to be believed. At the end of the sixteenth century and in the first decades of the seventeenth century the Icelandic scholar Arngrímur Jónsson wrote commentaries at the behest of Bishop Guðbrandur Þorláksson refuting earlier foreign descriptions of Iceland that had proved popular in Europe. His *Brevis commentarius de Islandia* of 1593 specifically challenged the work of the Hamburg merchant Gories Peerse, who had perhaps overegged the exoticism of Iceland in his poem that described its natural phenomena and cultural practices, suggesting that 'it would be truly fitting if he were to be devoured by lions.' In 1613 Arngrímur Jónsson's *Anatome Blefkeniana* countered the German preacher

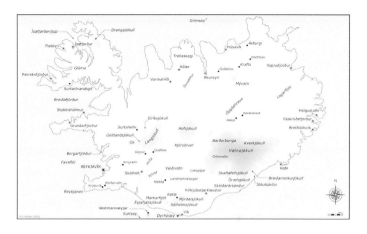

Map of Iceland with locations of places mentioned in the text.

Dithmar Blefken's 1607 *Islandia*, said to have been written following a visit to Iceland with Hamburg merchants. *Islandia* gives a less than flattering view of the country and its people and includes many falsehoods; Blefken wrote that he had met a two-hundred-year-old man and that Icelanders could live to the age of three hundred. Whether Blefken ever visited the island is still a matter of dispute and largely considered unlikely, a seventeenth-century example of the power of fake news underwritten by intriguing stories.[10] Despite these attempts to refute the falsehoods that were circulating in mainland Europe, by 1767 the French traveller Yves Joseph de Kerguelen-Trémarec still reported fallaciously the existence of marble, copper and iron mines in Iceland.

The British contribution to travellers' accounts in the eighteenth and nineteenth centuries comprised largely the reminiscences of a number of titled gentlemen, scientists, mountaineers and seasoned travellers, all of whom had the necessary wealth to support their travels. In 1772 Sir Joseph Banks, whom the king of Denmark accepted was travelling as one of a party of 'celebrated English Lords', was accompanied by artists, scientists, ten servants and two French horns.[11] Originally Banks was an accidental Icelandic explorer; he had been due to travel to the South Seas on the *Resolution* with Captain James Cook to search for 'Terra Australis', but he withdrew following criticism from the Admiralty that his group was too large and in danger of taking over the expedition. At a loose end and with his party all prepared, he turned his attention to the north.

Others chose to travel more conservatively. In 1809 the English botanist William Jackson Hooker travelled on the barque *Margaret and Ann* with a much smaller but no less interesting group, comprising a Danish seaman and a rather unscrupulous British merchant called Samuel Phelps, whose interests did not go further than politics and possible commercial opportunities. The Danish seaman turned out to be Jørgen Jørgensen, who for a brief time during that summer proclaimed himself king of Iceland, liberating Iceland from Danish rule before the Danish administration was restored with the help of the British navy in the form of HMS *Talbot*.[12] Hooker did have one of the most extravagant meals recorded by any visitor around this time – with ex-Governor Ólafur Stephensen, a friend of Sir Joseph Banks, at Stephensen's home on Viðey; this proved to be an eventful visit for Hooker, who also lost the majority of his papers and engravings in a fire on his return voyage to Britain. Yet despite the interest in the human side of what such visitors saw it is the landscape that dominates, its strangeness and its lack of conformity with anything that could be found at home, despite being only a few days' sailing to the west. Even someone like the Scot Ebenezer Henderson, in 1814–15, whose principal purpose in visiting Iceland was to propagate the Gospel and dispense Bibles on behalf of the British and Foreign Bible Society, commented:

> The very prominent place which the natural appearances of the island occupy on almost every page, arises from the predominance and extraordinary character of these phenomena. In no quarter of the globe do we find crowded within the same extent of surface such a number of ignivominous mountains, so many boiling springs, or such immense tracts of lavas, as have arrested the attention of the traveller.[13]

Tragically, for some strange reason 'ignivominous' was never adopted as a geological term, unlike several Icelandic words such as *sandur* and *þúfur* that now form part of the geological lexicon. Although fascinated by the science all around him, Henderson saw the phenomena as manifestations of divine omnipotence, frequently quoting passages from the Bible or his favoured authors and putting his life in God's hands as he crossed rivers. Daniel Vetter had similarly invoked the deity regularly throughout his account of his visit and attendance at the Icelandic parliament, or Alþingi, although his invocation arose mainly out of fear, especially after a sea crossing where his party encountered pirates, a severe storm and death on board, leading to one of the merchant's assistants being buried at sea.

To others, the landscape was seen as almost sacred but not in a religious way; for example, William Morris wanted to see the locations where the sagas were acted out and felt a special passion for Þingvellir, surprising the local Icelanders by his pilgrimages to see the homes of Njál

William Morris photographed in 1874 by Frederick Hollyer, just after his Iceland journeys in 1871 and 1873.

and Gunnar, with whom he was well acquainted through translations of the sagas. Morris was profoundly affected by his first Iceland visit; he is described on his return, standing at Edinburgh railway station, as 'short, fat, red-faced, bull necked, bushy-bearded and quite bewildered but very happy' – despite the knowledge that his wife Janey had been having an affair with the painter Dante Gabriel Rossetti in his absence. Morris in his turn had been influenced by the historian Thomas Carlyle and his description of Iceland as the 'waste chaotic battlefield of Frost and Fire', one of the earliest British writers to coin the idea of Iceland as a land of ice and fire.[14] Morris's fascination with the land of the sagas was echoed by other British travellers during the second half of the nineteenth century. George E. J. Powell, who with Eiríkur Magnússon translated Jón Árnason's *Icelandic Legends*, published in 1864, commissioned the German painter Johann Baptist Zwecker to paint a portrait of Fjallkonan, the Lady of the Mountain, the female personification of Iceland (Iceland's equivalent to the British Britannia).[15] The artist wrote to the poet and nationalist Jón Sigurðsson pointing out the crown of ice on her head, from which fires erupted, as symbolizing the essence of the country. Although black-and-white engravings can be seen widely, the original is actually kept in Wales at the University of Aberystwyth, as part of the Powell bequest. Reverence for the

Johann Baptist Zwecker, 'Fjallkonan (The Lady of the Mountain)', wood engraving copied by the artist from his original watercolour, published in *Icelandic Legends: Collected by Jón Árnason* (1866), translated into English by George E. J. Powell and Eiríkur Magnússon.

sagas did not always lead to respectful behaviour by the literary pilgrims. In 1897 the artist W. G. (Gershom) Collingwood undertook his *Pilgrimage to the Saga-steads of Iceland* with Jón Stefánsson and eagerly traced the footsteps of Guðrún Ósvífrsdóttir, the central female character in *Laxdæla saga*, to Helgufell in western Iceland.[16] Guðrún had four husbands, who were in turn divorced, drowned, killed and drowned, after which she turned to religion and became something of an anchorite. Her life was the focus

MOST UNIMAGINABLY STRANGE

of an important and much discussed poem by William Morris, *The Lovers of Gudrun*, his first Icelandic work that was not strictly a translation or retelling of a saga. Collingwood and Jón Stefánsson dug up her grave, then reinterred the body less some souvenirs, albeit in the name of academic curiosity. Interest in the more macabre aspects of the sagas and associated sites, coupled with the perils of accessing Iceland and travelling into the interior, have placed some of the nineteenth-century travellers as early practitioners of 'dark tourism' or 'thanatourism', visitors for whom a consideration of death lies at the heart of their desire to see and experience new locations.[17]

Not all those who visited Iceland fascinated by the sagas published accounts of their impressions of the landscape. The American bibliophile Willard Fiske (1831–1904), who bequeathed his impressive collection of Icelandic books to Cornell University (one of the three most extensive collections known, the others being in Reykjavík and Copenhagen),[18] travelled by horseback from Akureyri to Reykjavík looking at saga sites and seeking to acquire more books. He kept a diary but this is only available in the original. Fiske was, however, very well known in Iceland and feted everywhere he went. On his way to Iceland, as he was sailing to Akureyri, he passed close to the island of Grímsey and, learning about their enthusiasm for chess, arranged to give each family a chess set. For this he is now remembered every year by the island on his birthday, 11 November, with what the local literature calls, in their translation, an 'auspicious cake-buffet'.

In describing details of Fiske's journey after his death, Provost William Carpenter of Columbia University, who had been a student of Fiske and was introduced to Icelandic literature and language by his enthusiastic teaching, commented on the perceived difficulty for Americans working in a land 'as remote from our intellectual sympathy and understanding, as their island itself is remote from our ordinary journeying'. Americans began to visit Iceland in the middle of the nineteenth century, but with less of the intellectual and societal baggage of Europeans. Pliny Miles announced his arrival in Iceland in 1853 with 'the Yankee is here', taking more of a journalistic approach to his travels and seeking to rectify a perceived lack of information about the country.[19] By the end of the century travellers could not complain of a lack of information about Iceland as, in the words of the geologist Karl Grossman, who travelled across the country in 1889 and 1892, anyone spending time in the country seemed inevitably to succumb to an attack of '*furor scribendi*' (writing frenzy or mania) as he did himself.[20] Although most accounts of travelling across Iceland proved popular with audiences throughout Europe and America, there was still an underlying snobbishness in some literary circles about the country and those who visited, especially in Britain. In 1937, reviewing *Sheaves from Sagaland*, part of W. H. Auden and Louis MacNeice's *Letters from Iceland*, Sidney Barrington Gates talked of the poets following a 'long line of distinguished – or at least eccentric – predecessors in plodding across Iceland'.[21]

The sense of otherworldliness, described by the travel writer James (now Jan) Morris as 'a queer combination of the frigid and the lush, almost the erotic, which is the fascination of the place', persists to the present, as film-makers looking for a backdrop that is 'out of this world' choose to use various parts of Iceland for empty, desolate planetary landscapes, or fantasy lands frequented by Batman and comparable superhero-type characters.[22] There is now even an app to assist in identifying film locations when driving around Iceland.[23] In the BBC series *The Planets*, both Mars and the surface of Enceladus, an icy moon of Saturn, are portrayed using Icelandic backdrops. It is a favoured location for advertisers and music videos. In 2019 over sixty licences were requested for permission to film in the south of the Vatnajökull National Park (Vatnajökulsþjóðgarður) alone, 37 of these at Jökulsárlón, and 87 requests for the use of drones in the same area. The Grammy-winning Bon Iver video for their song 'Holocene' could have been made by the Icelandic Tourist Board (Ferðamálastofa), given its coverage of a number of the most popular and distinctive locations in the country, and both the title and the lyrics about seeing for miles are particularly apt.[24] Not only is Iceland different but it is seen as a constantly

Roni Horn, *Vatnasafn/Library of Water* installation at Stykkishólmur, 2007. Each cylinder contains water from one of 24 Icelandic glaciers.

MOST UNIMAGINABLY STRANGE

changing landscape, something acknowledged by the American artist Roni Horn with her installation at Stykkishólmur entitled *Vatnasafn* or *Library of Water*, based on water from a range of Icelandic glaciers. As Horn has said, 'Iceland is always becoming what it will be, and what it will be is not a fixed thing either,'[25] a sentiment echoed by the Danish-Icelandic artist Ólafur Elíasson,[26] whose works and installations, as showcased recently in his 'In Real Life' exhibition, reflect an idea of Iceland constantly changing and becoming; truly 'Art produced at the geographical margins of existence'.[27] It is also difficult to go for long through any shopping mall or watch adverts in cinemas without spotting an Icelandic background, especially in those that try to sell cars or clothes.

People travelled to Iceland out of curiosity and viewed the country from varied backgrounds, not just those with a particular enthusiasm for geology. We are fortunate that many accounts provided detailed observations of a wide range of geological phenomena, even if they were not the main reason for visiting. Not all prospective expeditions made it all the way to Iceland, and not always due to the vagaries and hazards of travel across the Atlantic. In 1799 Henry, Lord Brougham and Vaux, only got as far as Ullapool, where he ran out of alcohol due to excessive hospitality from the local laird: 'here we have again been at the flesh-pots and shooting and drinking.'[28] While later Victorian travellers such as Lord Dufferin, whose book of his visit in 1856, *Letters from High Latitudes*, was very widely read and ran to eleven editions in Great Britain and at least a further twelve across the Atlantic,[29] may have concentrated largely, but not entirely, on his eating, drinking and the quality of the shooting, others such as Henry Holland (1788–1873), who accompanied Sir George Mackenzie on his expedition in 1810, took copious notes on weather, geology, the temperatures of hot springs and so on.[30] This sometimes led to long journals with rather repetitive observations as they made their slow way over the lava plains. By persisting and publishing – the 'tyranny of the travel book, the duty to publish' seen as something of a millstone by Henry Holland – they have left an illuminating legacy. The growing catalogue of accounts had usually been thoroughly dissected by travellers before they departed from home, and in some cases both text and figures were duplicated. Careful analysis of the account of the visit in 1810 by Sir George Mackenzie,[31] which was very widely read, reveals significant use of the journal of his younger fellow traveller Henry Holland, and in parts is pure plagiarism. It was only with the publication of Holland's journal by the Hakluyt Society (edited by Andrew Wawn) in 1987, after it had appeared in an Icelandic translation, that the extent of Mackenzie's debt to Holland was fully understood. Ebenezer Henderson was unusual

View across Breiðafjörður, an area characterized by numerous low islands and home to Eggert Ólafsson.

For visitors spending just a few days in Iceland the customary excursion is to see the Golden Circle of Þingvellir, Geysir and Gullfoss; more than 50 per cent of the 1.2 million visitors to Iceland in 2014 (a figure that had risen rapidly to over 2.3 million by 2018) visited these sites.[41] Earlier travellers followed a similar route, heading out for Þingvellir after taking in the delights of Reykjavík, where they were introduced to Icelandic manners and cuisine, then travelling on to Geysir, and in some cases the slopes of Mount Hekla, although Gullfoss is rarely if ever mentioned and was presumably not visited. Douglas Hill Scott, in his *Sportsman's and Tourist's Guide to Iceland*, noted that there was still no regular road to Gullfoss at the end of the nineteenth century, and that crossing the Hvítá required a ferry with horses swimming the river. In 1873 William Morris undertook a more challenging journey through the Central Highlands to Akureyri and back, including his diversion to the 'most unimaginably strange' Markarfljót. Ebenezer Henderson spent the winter in Iceland and thus managed to see more of the country than many later travellers, and in 1860 another clergyman, the Reverend Frederick Metcalfe, the 'Oxonian in Iceland', took the unusual route of not only travelling across the Highlands to Akureyri, but then carrying on to areas of the northwest by crossing the mountainous Tröllaskagi peninsula.[42]

The time available for travelling was usually severely limited, not only by the weather and the quality of the roads, but initially by the need to catch a returning yacht to prevent overwintering in Iceland. In the first half of the nineteenth century in particular travel was by yacht, with Sir Thomas Wilson visiting in 1813, Lord Stuart of Rothsay in 1831 and John Barrow on the *Flower of Yarrow* in 1834. Travel improved in 1856 with the first regular steamship route between Denmark and Iceland plied by the *Arcturus*, a 364-ton iron cargo-passenger vessel built on the Clyde in Scotland, which called in at Grangemouth to pick up passengers. There were between six and eight sailings a year in the summer, allowing a five-day stopover, hence the frequency of travellers on the Þingvellir–Geysir route that could be managed within the time. Steamships may have offered a more reliable service but did not always keep to their timetable. Metcalfe found that his sailing took place two weeks after the stated date, and similarly his return date bore little relationship to what had been advertised, although he was reassured to find that the *Arcturus* had a Danish captain and Scottish engineers. Another clergyman, the Reverend W. T. McCormick, who dedicated the account of his trip to Lord Dufferin, missed his sailing in 1891 by assuming it would depart at six in the evening rather than in the morning, and had to spend two weeks in Edinburgh waiting for the next sailing.[43] The cost of the trip without food was £5 each way and views varied over the comfort of the voyage. William Lord Watts in 1876 talked of the floating purgatory of the *Diana*, which William Morris had used on both his visits, and preferred to take his chance with the aptly named Captain Cockle on the *Buda*.[44]

William George Lock saw an opportunity for an Icelandic Baedecker given the growing number of tourists and in 1882 published his *Guide to Iceland: A Useful Handbook for Travellers and Sportsmen*, adding the area known broadly as Njál's Country to the standard tourist itinerary.[45] Lock, who self-published, noted that in response to increased visitor numbers the oldest hotel in Reykjavík had been renovated to include a billiard and smoking area for their overseas male guests as Iceland's reputation of having good shooting and fishing spread, particularly among the richer British visitors. Once Reykjavík was left behind, he saw the benefit of Iceland as offering the sportsman 'a fortnight's glorious Arab-life'. Apart from Lord Dufferin, other travellers enjoyed the sporting aspects of their visit: Commander Charles S. Forbes, RN, managed to shoot six hundred snipe, ptarmigan and plover in five days in 1859,[46] and Metcalfe ensured he had strong cases for transporting his gun and fishing rod across the wilds. By the end of the century Lock's guide had been superseded by Douglas Hill Scott's *Sportsman's and Tourist's Handbook to Iceland*, which was specially prepared for passengers on the Royal Mail Steamers of the United Steamship Company operating out of Leith.[47] A preface to a new edition of *A Girl's Ride in Iceland* written by Mrs Alec Tweedie (Ethel Brilliana Harley) in 1894, five years after her visit, made a point of recommending a new sport's outfitters in Reykjavík that sold English cartridges and salmon flies, and emphasized the excellence of the shooting, not just of birds, but of reindeer, whale and seals.[48] In the later nineteenth century, Leith, the Scottish port principally used by the steamers, could boast a number of merchants with products specifically designed for the Iceland traveller: a ladies' inflatable bath, specially prepared food hampers and odour-free leggings, to name but a few. By the end of the century the number of sailings had risen to over thirty, reaching fifty by the opening of the First World War.

The importance of the steamships for connecting Iceland to Europe and for travelling around Iceland can be seen from reading Pike Ward's diary of his travels in 1906 when organizing the transport of fish caught off Iceland to Great Britain. The number of losses of fishing boats that he encountered in a single summer spent in Iceland is also a salutary reminder of the dangers of the waters around Iceland and across the North Atlantic in general. The opportunity the steamships presented for spreading the Iceland experience was not seen by all as a welcome development. In his later years Sir Henry Holland complained of ignorant tourists who had to see quickly the 'correct' attractions, and the seasoned traveller and explorer Sir Richard

Burton, who visited Iceland in 1872, coined the phrase 'Cockney excursions' for those rushing off to see Great Geysir.[49] Burton was extremely dismissive of the Iceland 'experience', feeling that everyone who visited had 'a distorted and exaggerated mental picture of what has not met, and will not meet, the eye of sense', and considered Iceland to have only 'humble features', with Hekla and Geysir as 'gross humbugs'. As Andrew Wawn observed, though, when writing about William Morris, 'by the 1870s Iceland became, for the nineteenth-century trendy traveller, what the lower slope of the Himalayan mountains have become a century later for the mobile-phoned merchant banker.'[50] This may have been the case but Mrs Disney Leith, Mary Gosden, who made no fewer than fourteen trips between 1894 and 1914, still considered Iceland 'unsuited for anyone who is not in fairly robust health, a good sailor and able to enjoy riding'.[51] She was 54 on her first visit and climbed Mount Hekla with her daughter at the age of seventy.

The last steamship usually sailed in late August, or very early September, and there were spaces for 26 passengers, normally returning to Europe with between sixty and seventy Icelandic ponies destined for British coal mines. The *Camoens*, a Portuguese steamship, on which Mrs Alec Tweedie followed a rather unusual route from Granton in Edinburgh to Akureyri, Sauðárkrókur, Blönduós and finally Reykjavík, took on 617 ponies before returning to Scotland, seventeen of which died in passage and had to be thrown overboard. In 1888 the ss *Copeland* ran aground between Orkney and the Scottish mainland and 360 of the 482 Icelandic ponies on board managed to swim ashore; all the passengers – including the author H. Rider Haggard – were saved. Ponies were still being shipped in the twentieth century and in 1931 Alice Selby described her boat calling at Vík to pick up ponies which were packed tightly in the hold to prevent them from lying down and being trampled.[52] With the improvement in communications, news and information was able to travel back and forth from Iceland much more quickly than in previous centuries. Having completed his crossing of Vatnajökull in the summer of 1876, visiting areas of northern Iceland in the time before his steamship sailing, William Lord Watts was able to see a copy of the *Evening Echo* giving a 'deplorable account of his health and personal appearance following his challenging journey'. A small number of hardier travellers, including William Lock, stayed on over the winter, in particular Ebenezer Henderson, who in 1814–15 took his Gospel propagation to areas in the north, east and west rarely visited previously or reported on by foreigners. Like Ida Pfeiffer, who in 1845 probably slept in more churches than any other visitor before or since, Henderson stayed principally with ministers who were spread across the island overseeing the pastoral needs of the population. Later in the nineteenth century Sabine Baring-Gould, of the hymn 'Onward Christian Soldiers' fame and whose appreciation society has only recently been wound up, similarly preferred the 'comforts' of churches to the perceived questionable hygiene of most houses while travelling to Iceland between graduation and marriage. At Geysir he stayed in the church at Úthlíð and

> supped off coffee and Icelandic moss stewed in milk, which the farmer's wife laid for us on the communion-table. We ate it, lying on our beds placed within the rails, by the light of the altar candles, which the good woman had kindled to do us honour.[33]

Beds and food were usually available for travellers at farms whenever they appeared, but some came prepared for a potentially uncomfortable night. The Reverend W. T. McCormick took with him Keating's insect powder, which he spread on the floor before turning in. This was very popular and widely used against beetles, bugs, fleas, cockroaches and moths, with the instructions recommending that it should be sprinkled on beds and pillows, ensuring that all the crevices on the bed were treated. Its main active ingredient was pyrethrum and it was still in use in the trenches in the First World War to attempt to control lice.

The difficulties faced both by these travellers, especially in winter, and also by the local doctors and ministers who, while carrying out their duties, observed, made notes on and contemplated the landscape they saw around them, should not be underestimated. Crossing glacial rivers looms large in all accounts and was a dangerous, uncomfortable and frequent undertaking, as cautioned by Henderson:

> The fording of the rivers, the climbing of the mountains, the scrambling over the lava, the passage of the morasses, bad weather, and numerous other circumstances, present very serious inconveniences, even to the most robust and accustomed traveller, and might be deemed absolutely insurmounted barriers in the way of females and young children.

Ida Pfeiffer must only have seen this as a challenge and proved a much hardier and adventurous traveller than most of her male counterparts. Even in 1906, travelling to Nýpsfjörður in eastern Iceland, the Devonian Pike Ward, seen as a founder of the commercial Icelandic fishing industry and once termed the 'best known man in Iceland', had many issues crossing rivers:

Sprengisandur route across the Central Highlands with a view of Hofsjökull.

This is a dangerous river . . . Jakob goes on ahead and I follow, and it gets deeper and deeper until it reaches up to my knees and my boots get full of water . . . once these ponies begin to go over a river there is no stopping . . . so I cling on. We crawl up the bank to dry land and make for a grass flat. Get off and off-saddle and sit down and change our socks, empty the water out of our boots, burn pieces of paper inside to dry them.[54]

The largest rivers did have ferries but these did not always inspire confidence. McCormick described the ferry over the Þjórsá as the most dangerous he had ever seen, 'rotten but cobbled up like an old shoe . . . with water spouting in like small geysers'.

It is worth remembering that it was only in 1974 that the bridge was completed across Skeiðarársandur, providing the final link to road travel around the whole country. Travel over much of the country up to the Second World War, and later in the interior, meant time spent on horseback. Pfeiffer made it further up Mount Hekla on horseback than her guides had previously achieved without dismounting, so to travel successfully required good horse skills, or at least a docile mount. Given the title of her account, *A Girl's Ride in Iceland*, it comes as no surprise that significant parts of Tweedie's work concentrate on the quality and behaviour of her ponies. She scandalously dispensed with riding side saddle on her four-day 260-kilometre (160 mi.) round trip from Reykjavík to Geysir, a feature highlighted in reviews of her book, which went into several editions, detailing her 'plucky' exploits.[55] Riding astride the horse was seen by Barrow and McCormick as something restricted to poorer women, who could not afford the local wooden side saddles described by Morris as 'little chairs with gay-coloured pretty home-woven carpets thrown over them'.

As the twentieth century progressed, four-wheel-drive improvements to cars, trucks and coaches made areas of the interior more accessible, first for Icelanders and then increasingly for foreign tourists. On his way to Veiðivötn in 1950 Guðmundur Jónasson discovered the ford over the Tungnaá, so opening up Sprengisandur from the south, and along with Úlfar Jacobsen, ensured that the Central Highlands became a feasible tourist destination, albeit only during the summer and not in great comfort.[56] When travelling from Reykavík to Núpsstaðir by bus, lorry and horse, to begin an expedition to the western part of Vatnajökull in 1935, Andrea de Pollitzer-Pollenghi recalled the travails of crossing the Múlakvísl *sandur* plain:

I really cannot imagine a worse road than the one we were to be
driven over. We were proceeding along a track; every moment we
had to stop and make a fresh start. We had to get down and push
the car, run after it, perspiring in our heavy black oilskins and
rubber boots.[57]

In 1936 the glaciologist Hans Wilhelmsson Ahlmann still had to use
horses to transport equipment to his base and onto the glacier for the
first detailed studies undertaken on the Vatnajökull ice cap as part of the
Swedish-Icelandic expedition to Vatnajökull.[58] Almost twenty years later,
on the ill-fated 1953 Nottingham University Iceland Expedition to southern
Vatnajökull, during which two members died high up on the crevassed ice
above Skaftafellsjökull, tractors were used to get the supplies and scientific
equipment across the same rivers.[59]

    The challenges of the physical environment meant that it was not
until the period between 1900 and 1939 that the Danish Geodetic Survey
produced maps of Iceland at a scale of 1:50,000, with total coverage at
1:100,000 achieved by independence in 1944. For the earliest visitors, the
1590 map of the island by the Flemish cartographer Abraham Ortelius,
which was variously revised and was a considerable improvement on
previous maps such as that by Benedetto Bordone in 1528, proved the only
reliable source.[60] While stylized and best known for the array of different
sea creatures shown in the ocean around the coast, it included a number
of accurately portrayed features and settlements, especially the bishoprics
of Skálholt and Hólar, reinforcing the view that the map owed much, if
not all, to earlier mapping by Guðbrandur Þorláksson, Bishop of Hólar.
In 1752 Niels Horrebow's *Natural History of Iceland* was accompanied
by a map produced by the Norwegian Thomas Hans Henrik Knoff in
1734 at the request of the Danish government.[61] Because Knoff had sent
some of his early maps to his Norwegian superiors, actions considered
unacceptable in Denmark, the Danish king forbade publication of his
work and it was only in 1752 that it was published. It was an improvement
on Ortelius but still included some errors and fantasies. Between 1831 and
1843 an Icelandic teacher, Björn Gunnlaugsson, managed to survey most
of the country, including much of the uninhabited interior, providing the
first reliable outline of lava fields, glaciers and rivers. His surveys were
of very high quality despite his view that users 'should neither have too
high or too low expectations of the map, nor trust too greatly, nor too
little in its usefulness or accuracy',[62] and provided the basis for the maps
used by the increasing number of travellers in the second half of the
nineteenth century. This included the tourist map produced by Edward

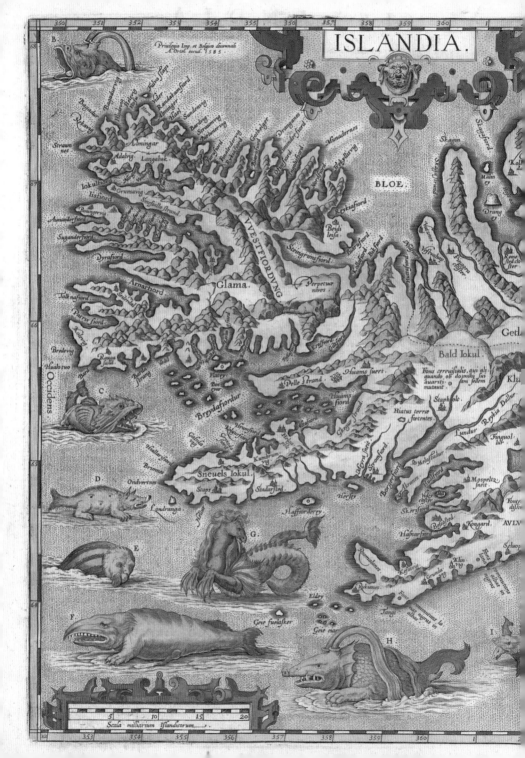

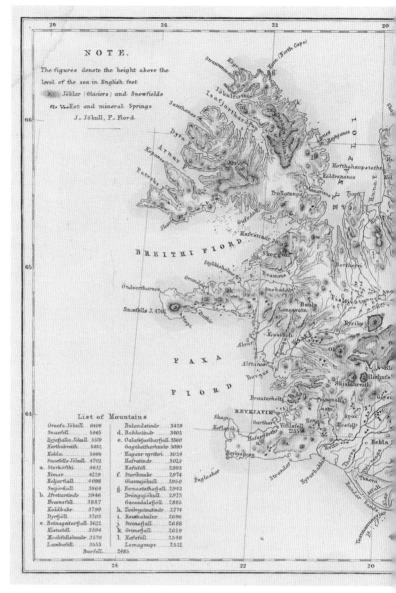

G. H. Swanston, after Augustus Petermann, 'Iceland According to the Trigon Survey of [Björn] Gunnlaugsson, 1844', map published in Archibald Fullarton, *The Royal Illustrated Atlas of Modern Geography* (1862).

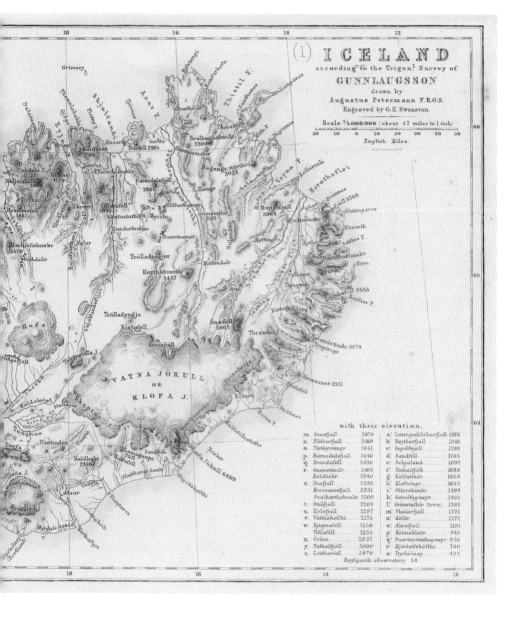

ICELAND

according to the Trigon.l Survey of

GUNNLAUGSSON

drawn by

Augustus Petermann F.R.G.S.

Engraved by G.H.Swanston

Scale 1/3,000,000 (about 47 miles to 1 inch)

20   10   0   10   20   50   40   50

English Miles.

with their elevation.

m.	Snaefiall	2478
n.	Hlitharfiall	2469
o.	Tarihyrnunqr	2451
p.	Barnadalsfiall	2440
q.	Brandafell	2436
r.	Gaesartindr	2405
	Kaldbakr	2340
s.	Diasfiall	2333
	Krossanesfiall	2331
	Svalbardtshnukr	2300
t.	Stalfiall	2203
u.	Holsfiall	2197
v.	Vathlakeilhu	2175
w.	Rjupnafell	2159
	Volufell	2185
x.	Orkn	2037
y.	Vathalfiall	2000
z.	Leitharool	1979

a.	Launquhltiharfiall	1978
b.	Reytharfiall	1946
c.	Ingolfsjiall	1789
d.	Sandfell	1705
e.	Selgaland	1694
f.	Vathalfioll	1680
g.	Kalfatindr	1658
h.	Klofningr	1642
i.	Ollarshnukr	1494
k.	Geirolfsgnupr	1465
l.	Grimstathir (town)	1393
m.	Skalarfiall	1375
n.	Keilir	1272
o.	Akrafiall	1191
p.	Heimakleitr	940
q.	Snartarstathagnupr	926
r.	Hjorleifshofthi	760
s.	Dyrholaey	402

Reykjavik observatory 58

Weller, printed between the 1850s and 1880s and published in monthly instalments in *Cassell's Weekly Dispatch Atlas*, part of the British *Weekly Dispatch* newspaper.[63] Björn Gunnlaugsson's surveys were made into maps by Olaf Nikolas Olsen (1794–1848), head of the topographic department on the Danish General Staff in Copenhagen, who suggested producing it in four sheets. This map appeared in 1844 and was the one purchased for two guineas by Metcalfe at Stanfords in London prior to his visit in 1860.

Despite the long literary tradition of the Icelanders, relatively little of broader geological interest can be found in the saga literature. There is an excellent eye for local topographic detail, such that farms and locations mentioned in various sagas can be reliably located, and this facility for revisiting the great sites such as the farmstead at Bergþórshvoll, burned to the ground in *Njáls saga* (*Brennu-Njáls saga*), provided a particular draw for British saga-lovers including William Morris, Sabine Baring-Gould and the artist W. G. Collingwood. Annals and various other early documents recount events such as floods, volcanic eruptions and extreme weather, especially the *Eldrit*, which in a rather matter-of-fact way reported on volcanic eruptions, their nature and effects. Only recently has a systematic attempt been made to survey the extant Icelandic literature between circa 800 CE and 1800 to look for records of the natural world – by the ICECHANGE project of the Stefansson Arctic Institute, an interdisciplinary institute set up in 1998 at Akureyri in northern Iceland.[64] A highly descriptive environmentally grounded approach can be seen, though, in Icelandic literature, and while not scientifically questioning like many of the visitors' accounts, numerous novels provide a unique insight into the reality of living with the challenging landscape and weather. From the details of the daily life of Bjartur at the isolated ill-named farm Summerhouses (*Súmarhúsum*) in the Nobel Laureate Halldór Laxness's novel *Independent People* (*Sjálfstætt fólk*),[65] to the trials and tribulations of 'the boy' in Jón Kalman Stefánsson's award-winning trilogy that includes *Heaven and Hell* (*Himnaríki og helvíti*), *The Sorrow of Angels* (*Harmur englanna*) and *The Heart of Man* (*Hjarta mannsins*), set in western Iceland in the early 1900s, there is a visceral feeling of the nature of the physical environment. With Jón Kalman Stefánsson this is particularly noticeable in *The Sorrow of Angels*, where 'the boy' accompanies Jens the postman on his round across fjords and mountains in a chapter entitled: 'The Journey: If the Devil Has Created Anything in This World, Besides Money, It's Blowing Snow in the Mountains'.[66]

Ideas of Iceland held by visitors and prospective visitors tend to focus on iconic landscapes, which represent the different aspects of how the overall landscape developed. The following chapters seek to use some of these as a basis for a wider appreciation of the processes that have created the landscape we see today, and which in many cases differ little from those seen by visitors centuries earlier. In providing an overview of the geology and landscape of Iceland, the opportunity is taken to look in a little depth at the variety of individuals, both foreign and Icelandic, who over the last millennium have been intrigued and challenged by the strange land lying at the very edge of imagination, and whose observations and thoughts can add much to our understanding of the landscape even today. From Uno von Troil, later to become Archbishop of Uppsala, whose observations from the mid-eighteenth century proved essential reading for anyone visiting Iceland, to W. H. Auden and Louis McNeice with their *Letters from Iceland* homage to their visit in 1936,[67] revisited almost sixty years later in *Moon Country* by the poets Simon Armitage and Glyn Maxwell,[68] writers all contribute in their own ways and help explain an enduring fascination that goes well beyond grasping the simple geological truths according to contemporary theories. In an interesting reference in the 'Sheaves from Sagaland' chapter of *Letters*, Auden and MacNeice quote Pliny Miles, who visited the country in 1853, saying, 'Iceland is not a myth, it is a solid portion of the earth's surface.'[69] Travellers may have been attracted to Iceland by myths, sagas and tales of the exotic landscape, but once there it is very much the stark reality of the country as evoked in its landscape that provides the key to their reactions and impressions. As far as Icelandic observers are concerned, standing outside his church in Kirkjubæjarklaustur in 1783 praying for the Laki lava to stop, the cleric Jón Steingrímsson may have had little grasp of the real nature of the geological environment in which his community was living, but he had a tremendous eye for detail and an awareness of natural processes, qualities found widely in the literate society of Icelanders.[70]

Images of Iceland presented outside the country still relied heavily on the written word well into the twentieth century. Before the advent of the camera, accounts were supplemented by engravings, done either by authors or fellow travellers taken along because of their artistic skills. In 1862 Sabine Baring-Gould made a point of championing the sketches and paintings he produced in his account of his travels, but overall where engravings or images were produced by visitors they tended to range in quality and verisimilitude, often exaggerating the scale and immediacy of what was being represented. The Irish-born American traveller J. Ross Browne, whose account of his journeys across Russia and Scandinavia, *The Land of Thor*, published first in 1867, became a very popular travelogue at its time, bemoaned the lack of images he was able to find of Iceland,[71] yet at almost exactly the same time A. J. Symington, another enthusiastic saga lover, produced his *Pen and Pencil Sketches of Faroe and Iceland* (1862).[72]

Ross Browne decided to make a brief visit on the *Arcturus* and vowed to rectify the omission with his own work. Because he was running out of money on his travels, he had considered going to Iceland to look for a specimen of the great auk to sell to the British Museum, but decided this was too risky. This was a wise decision as the last known pair were killed on Eldey in 1844. Accompanied by Geir Zoëga, 'a jewel of a guide who knows every rock, bog, and mud-puddle between Reykjavík and the Geysers . . . born in all probability of an iceberg and a volcano',[73] he visited Þingvellir and Geysir and produced a number of sketches as he had promised, not without occasional exaggeration, but generally very accurately. Ross Browne was not being totally fair to his predecessors, not just his contemporaries, in that Sir Joseph Banks had three artists with him in 1772, John Cleveley and the brothers James and John Frederick Miller. They produced a number of what have been described as 'finely observed' drawings and paintings, some of which were exhibited on their return, even if they did not appear in a published journal of the travels as might have been expected. Four volumes of these sketches and drawings were bequeathed to the British Museum.

Because of the otherness and exoticism of Iceland, in a publication produced to accompany the first exhibition of Icelandic art in England in 1989 John Russell Taylor argued:

> on-the-spot travellers contrived, through even the dullest and most scientific prose and then through graphic depictions which, even if they began as topographic record, ended as Sublime and apocalyptic visions, to print a remote and magical image on the inner eye of the armchair traveller too, and on the consciousness of the general public.[74]

This comment does not do justice to W. G. Collingwood, who produced over three hundred paintings and ink sketches from his visit to the country – which he named 'Niceland' – tracing the saga-steads; many of these are now held in Iceland at the National Museum (*Þjóðminjasafn Íslands*). Collingwood made a point of producing monochromatic images using ink in an attempt to mimic photographs, to add a greater perceived accuracy to his reproductions of the key saga landscapes. Although he was not averse to occasional exaggeration in the scale of the mountains, Collingwood's pictures are reliable reproductions of what he saw, as pointed out by Matthias Egeler in his recent publication of a facsimile copy of *Saga-steads* in 2015, an observation he reinforces by adding colour photographs of Drangey and Surtshellir alongside the original figures. Images of any kind were of

W. G. Collingwood, 'Hvamm in Hvammssveit', illustration of the Iceland mountain landscape from W. G. Collingwood and Jón Stefánsson, *A Pilgrimage to the Saga-steads of Iceland* (1899).

importance for use as accompaniments to lectures, not only by those who had visited and written about Iceland, but by others eager to educate the general public about the world beyond their immediate surroundings. The Yorkshire-born explorer and surgeon Tempest Anderson (1846–1913) had a particular interest in volcanoes and travelled extensively, including two visits to Iceland in 1890 and 1893, taking images that he then used in magic lantern shows on his return to England.[75] Not all lecturers, though, needed images to enable them to paint a compelling picture. In America in the middle of the nineteenth century the pre-eminent practitioner in this field was Bayard Taylor (1825–1878), whose theatrical lectures were given to overflowing crowds, including the president, Abraham Lincoln. He had celebrity status and was even stalked by some lady admirers.[76] He combined visits to Egypt and Iceland in 1874, and the perils and exoticism of Iceland, despite it being only a 'slightly out-of-the-way region',[77] appealed to his audiences, who could also purchase his *Cyclopaedia of Modern Travel*, a compilation of edited versions of travellers' accounts, including that of Ida Pfeiffer.

Some later nineteenth- and early twentieth-century publications are complemented by photographs, as in Watts's crossing of Vatnajökull, and Cornell University's Fiske Icelandic Collection now holds an important database of Icelandic photographs, especially those taken by Frederick W. W. Howell at the very end of the nineteenth century.[78] These include pictures of both volcanoes and glaciers that were advertised for sale in Douglas Hill Scott's *Sportsman's and Tourist's Handbook to Iceland*, but they were unlikely to have had a very wide distribution, and the power of the word and the images it elicited would still have probably been the main source of external impressions of Iceland until the latter half of the twentieth century. Photographs taken by the Dutch photojournalist Willem van de Poll on his tour around Iceland in the summer of 1934, only recently unearthed, provide a unique snapshot of the country before the Second World War but are mainly of people, with few of the landscape.[79] There are, though, several images of cars traversing very rough roads, including one in a river crossing, one at Gullfoss and another where it had missed the track and ended up on its side, providing a talking point for several local people reviewing the accident in the photograph. He was also not averse to enhancing his pictures with an occasional attractive woman, as in his image of ropey lava with a well-dressed female leg for scale and possibly the same young lady, rather more scantily clad, hand cranking the engine of one of the cars he was fond of showing. This was perhaps to be expected given that one of his many contemporary photographic roles was as a fashion photographer working for *Vogue* and *Harper's Bazaar*.

Indigenous Icelandic art did not really have a landscape tradition until the end of the nineteenth century with the work in particular of Ásgrímur Jónsson (1876–1958) and eventually the internationally recognized artist Jóhannes Sveinsson Kjarval (1885–1972), whose early paintings soon took on extra mythical dream-like features, giving the landscape an imaginative dimension.[80] A lot of Jóhannes Sveinsson Kjarval's work is deeply rooted in the physical form of the land, beautifully portrayed in what appears to be either a minimum of strokes, as in *Esja 10. februar 1959*, or in a very detailed representation of the surface, as in his paintings of lavafields, or *hraun*, and of rocky surfaces – such as *Skjaldbreiður (Í Grafningi)*, painted between 1957 and 1962. Jóhannes Sveinsson Kjarval was considered to be something of a typical bohemian artist, but as with Ásgrímur Jónsson, from whom he learnt as a young man, his origins were in farming, just like most other Icelandic scholars and artists. Despite the relative popularity of the watercolours by Baring-Gould and Collingwood, in discussing contrasts between Icelandic and 'English' attitudes to painting the Icelandic landscape, Halldór Björn Runólfsson suggested:

> Nothing is more alien to our rough temperament than the medium of watercolour. A medium which was admirably suited to the needs of cultivated Englishmen during the second half of the eighteenth century has never agreed with artists from such an unsophisticated farming culture as ours.[81]

Jóhannes Sveinsson Kjarval,
Mount Esja, 10. februar 1959,
oil on canvas.

Notwithstanding its unique character, relatively few non-Icelandic artists have been attracted to the challenge of representing the landscape on canvas. One English landscape artist of renown, Eric Ravilious (1903–1942), desperately wanted to paint in Iceland, a country he told his wife Tirzah that he had to visit as it was 'the promised land'; he spent only a short time in the country when posted there during the Second World War and tragically did not produce any work to extend his exceptional repertoire.[82] Arriving in August 1942 he was immediately posted to Kaldaðarnes in eastern Iceland and in less than a week went missing when his reconnaissance plane was lost in a search for another missing plane. In whichever way the land has been portrayed, the landscape of Iceland appears always to be associated with something more than that which can simply be seen, what the Icelandic poet Jóhann Hjálmarsson has described as 'the idea of a land'.[83]

Whatever medium was used over the years, the very idea of Iceland was rooted in its landscape and history and their apparent indivisibility. The images conjured up, though, said as much about the origins, culture and outlook of the observer as about the land itself. Sitting at his breakfast table in Oxford in early July 1860 'Oxonian in Iceland' Metcalfe perused adverts in the *Times* and saw that the *Arcturus* was to depart from Leith for Iceland in a week's time. Encouraged by his wife and driven by the excitement of the sagas, as well as an awareness of current fundamental evolutionary and geological debates, he decided to see the country for himself. This decision must have been taken with some trepidation because his image of Iceland, based on what he could find written by earlier travellers, was summarized as:

> by no means cheering . . . a people . . . abiding for the most part in dark caverns of dwellings, anything but attractive to a Briton, and so sparse in number that you must traverse three square miles . . . to find a couple of them. While, as if to mock the foreigner for his infatuation, his way is beset, not only by dangerous rivers, appalling lava streams, hidden pits of fire, and chasms of ice, but the imagination is tortured by chimeras dire, phantom gorillas, or by whatever he may please to call the shapes in stone and slag, that grin and frown at him on his solitary journey.

Surprisingly he quite enjoyed his visit, despite having an initial view that saw Iceland in the terms encapsulated by Benedict Gröndal as 'terra incognita et barbara'.[84]

This image can be contrasted with that produced in *A Report on the Resources of Iceland and Greenland*, written for the United States State Department in 1868 by Benjamin Mills Pierce, although produced by one of his staff. Neither Pierce, nor the R. J. Walker who did the work, had visited the country, but based their synopsis on the same literature seen by Metcalfe. Rather than just satisfying intellectual curiosity as in Metcalfe's case, there was a political dimension to the report as Pierce was attempting to persuade the Secretary of State, William H. Seward, to put the case to the government for the United States to purchase Greenland and Iceland, following Denmark's recent sale of the islands of St Thomas and St John. This would be a strategic move to hinder British naval power but also to look for economic opportunities. On the basis of all the available evidence Pierce (Walker) writes that

> It has fields beautifully green, mountains clothed in purple heath, and the atmosphere of astonishing purity. The lava in time becomes soil and pasture-land. Much of the heath can be made pasture-land . . . in view of its pastures and arable lands, its valuable mines, its splendid fisheries, and its unsurpassed hydraulic power, it could . . . sustain a population exceeding 1,000,000 . . . sulphur mines are very rich and extensive, easily worked and of immense value . . . etc.[85]

The contrast between Old and New World views is stark and manages to reflect the complete spectrum of reasons for visiting Iceland that can be encountered in the written accounts. Yet underlying both is the importance of the natural landscape in all its fascination and awfulness.

Scientists studying the geology and landscape of Iceland, especially over the last century, have similarly recognized that a more satisfying understanding of the physical nature of the land goes beyond mere observation and theory. This is best summed up by Hans Wilhelmsson Ahlmann, who not only undertook a joint expedition with Icelandic scientists but spent time with local farmers and explored beyond the confines of his glaciological objectives. In describing his time in Iceland and considering his own work he concluded:

> Natural science studies have predominated . . . [but] . . . usually give only facts and results of a special kind, but not a picture of the whole, or the soul of that island in which human beings live a life of their own, cut off from the world.[86]

As Philip Marsden wrote about the voyaging St Brendan, who must have sighted Iceland on his epic voyage across the Atlantic in the sixth century CE, 'Remain innocent, always open to the world's marvels. And don't go throwing books of wonder on the fire.'[87] In the spirit of Ahlmann, the aim here is to go beyond simple description and explanation in exploring the geological record, that which Charles Darwin called 'a history of the world imperfectly kept'.[88] Visiting Iceland is very much about the experience of being in a 'different' environment, somewhere with an adventurous overtone, an 'unimaginably strange' place that has never ceased to attract travellers, artists, writers and scientists over the centuries. *Þetta er Ísland* – such is Iceland.

CHAPTER 1

ÞINGVELLIR, THE UBIQUITY OF BASALT AND THE MAKING OF ICELAND

'EVERYTHING IS OF LAVA.'[1]

'LAVA, LAVA, LAVA IS THE ETERNAL VISTA.'[2]

Seen from the perspective of the early twenty-first century, with our knowledge of the Earth comprising a series of moving plates being created and destroyed over geological time, Iceland, as a place where processes of crustal creation can be observed occurring at the surface, offers a unique experience. Þingvellir epitomizes the whole idea of a moving Earth where one can stand astride the plates and imagine being stretched in opposing directions – given sufficient geological time, of course. For earlier travellers who knew of volcanoes and had either seen or heard about lava, visiting Þingvellir may not have been central to their understanding of how the planet evolved, but it still left long-lasting impressions and was the 'go to' place for anyone who had made the considerable effort to get to the country by sea:

> I was arrested in full career by a tremendous precipice, or rather chasm, which suddenly gaped beneath my feet . . . I was never so completely taken by surprise. We had reached the famous Almanna Gja. Gazing down with astonished eyes over the panorama of land and water embedded at my feet I could scarcely speak for pleasure and surprise . . . At last I have seen the famous Geysers, of which everyone has heard so much; but I have also seen Thingvalla, of which no one has heard anything . . . more wonderful, more marvellous is Thingvalla.[3]

When not shooting or fishing and generally enjoying the culinary delights from his sport, Lord Dufferin, more correctly Frederick Hamilton-Temple-Blackwood, Marquess of Dufferin and Alva, was keen to describe his very genuine feelings of amazement and curiosity

Frederick W. W. Howell,
photograph of Þingvellir,
c. 1900.

aroused while travelling around Iceland. His *Letters from High Latitudes*,
describing his voyage in the schooner *Foam* during which Iceland was
the first stopping-off point, before heading further north to Jan Mayen
and Spitzbergen (Svalbard), includes several drawings providing faithful
reproductions of the landscape. This was unlike many others who chose to
produce composites of key features such as Þingvellir, Hekla and glaciers,
nicely grouped together with locals or horses for scale. He also included a
small number of simple plans and schematic cross-sections to show how
he thought features such as Geysir may have operated or Þingvellir formed.
Although in many ways a classic Victorian aristocrat – with a peerage,
estates in Ireland and eventually a successful career within the British civil
service that saw him serving the empire as Viceroy of India and Governor
General of Canada among other posts – he was, like others, intrigued by
the Icelandic landscape. He may only have had a couple of years at Oxford
and left with what a later Master of University College described as a
'humble pass', but he demonstrably had a love of all things north and an
ability to convey the underlying beauty and otherworldliness of what he
saw. Descending the gorge at Almannagjá he details his feelings:

> So unchanged, so recent seemed the vestiges of this convulsion, that
> I felt I had been admitted to witness one of nature's grandest and
> most violent operations, almost in the very act of its execution . . .
> In the foreground lay huge masses of rock and lava, tossed about
> like the ruins of the world.

Þingvellir, looking northeast over the graben.

ENTRANCE TO THE ALMANNAGJA.

The idea of a violent land still undergoing change, in which lava was dominant and ubiquitous, is a feature of travellers' accounts and is still one of the attractions for visitors. Europeans knew of volcanoes and lava from experiences or descriptions of Mediterranean examples. Ida Pfeiffer had climbed Vesuvius and put her knowledge to good use in Iceland; trade and travel around the world through the nineteenth century began to make literate societies aware of more exotic 'active' locations with volcanic histories, as in the East Indies. These distant locations only served to contrast with the quiet, ancient and very stable landscapes in Europe from which most travellers originated, hence their otherworldliness and fascination. The ubiquity of lava is apparent to anyone traversing the country. Ninety-seven per cent of visitors now arrive by air at Keflavík and have to cross a series of lava flows from one of the most geologically active areas of the country, whether heading for Reykjavík or further inland. Ten per cent of the surface of Iceland has been formed in the last 11,700 years (the Holocene Epoch,[4] our current interglacial) – an area of around 10,000 km^2 (3,900 mi.^{2}) and a volume of 400 km^3 (96 mi.^{3}). Of this, 90 per cent is basalt.

Just how lavas formed and their role in earth history were questions that during the nineteenth century became a focus of much geological debate. Indeed, this debate was not satisfactorily resolved until the recognition of continental drift, and subsequently of plate tectonics over a century after Lord Dufferin's experiences. What Ida Pfeiffer in particular contributed was an appreciation of just how variable lava could be:

> Everything is of lava . . . like waves in a petrified sea.

> The manifold forms and varied outlines of the lava-fields present a remarkable and truly marvellous appearance.

> the lava streams seem to have been called into existence by magic, as there was no mountain to be seen . . . from which it could have emerged.

> masses of lava . . . of singular beauty . . . black mounds, ten or twelve feet in height, piled upon each other in the most varied form . . . while the tops were broken into peaks and cones of the most fantastic shape.

> The lava here assumes a new character . . . in the form of immense slabs or blocks of rock, often split in a vertical direction . . . masses of lava formed beautiful groups, bearing a great resemblance to ruins of ancient buildings.

> The most beautiful masses of lava, in the most varied and pictur-esque form . . . forming exquisite flames and arabesques.

> All lava is not the same; there is jagged, glassy and porous lava; the former is black . . . the farther the lava and sand from the mountain, the more they lose this blackness.[5]

In general, she found the lava landscapes fulfilled all her anticipated hopes, but she was not always so enthused and at times described 'gloomy monuments of volcanic agency'; 'new scenes of deserted, melancholy districts were revealed to us; everything was cold and dead, every where was black burst lava.' Nevertheless, her descriptions provide an insight into the variations of lava that she came across, reflecting differences in their ages and modes of formation.

We now distinguish two broad groups of volcanic rocks: lava originating from fissures and vents, in contrast to that erupted explosively and fragmented during eruption into different sizes and shapes of particles from bombs to ash. The separation of the former into *pahoehoe* or ropey lava (in Icelandic *helluhraun*) and *aa* lava (*apalhraun*) refers to the surface texture of the lava, linked to its viscosity, which in turn is associated with flow temperature, rate of flow and the slope of the terrain over which it flows. The very fluid *pahoehoe* lava spreads out as thin and often extensive sheets,[6] whereas the *aa* lava forms much thicker blocky flows. Lava in a *pahoehoe* flow is protected by the crustal surface, which quickly cools on contact with the air, allowing the rate of flow to slow and the formation of lobes that are then broken through by new hot lava, or toes of lava. The construction of the insulating surface and injection of more lava can lift up the surface of the flow, sometimes by tens of metres, a process known as lava inflation. The characteristic surface of *pahoehoe* lava was very neatly described by von Troil from his visit in 1772: 'the crust . . . generally forms a resemblance of a rope or cable.' He also identified how lava tunnels or tubes could form in *pahoehoe* lava, as the upper cooled crust protected lava below from cooling and hence allowed it to flow within the tunnels, although two decades earlier Eggert Ólafsson and Bjarni Pálsson had

Pahoehoe lava (*helluhraun*) at Þingvellir.

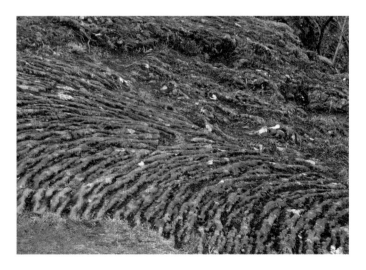

MOST UNIMAGINABLY STRANGE

written about the same phenomenon, based on their visit to the lava cave at Surtshellir to the north and inland from Reykjavík. Eggert Ólafsson had previously explored the cave in 1750 but was not well enough prepared to see much of its 1.6 km (1 mi.) extent. Details of the cave were first published in Danish and only translated into German around the time of von Troil's visit. These included detailed descriptions of lava stalactites and stalagmites – now termed lavacicles – along with measurements, and correctly suggested that these lavacicles were formed contemporaneously with the opening of the inflated lava tube, while the lava was still active. Eggert Ólafsson was another Icelander who achieved a lot in a very short life, despite being born on what would have been the relatively isolated island of Svefneyjar in Breiðafjörður, northwest Iceland. He died in 1758 aged 32, shortly after completing his travels, when he drowned with his wife on their honeymoon while crossing Breiðafjörður.

W. G. Collingwood's faithful depiction of the surface of the cave at Surtshellir in 1897 showed a number of visitors descending into the opening, to explore the lava tunnels, and the Raufarhólshellir lava tube east of the Bláfjöll mountains near Reykjavík is now a fast-growing tourist experience. The Reverend W. T. McCormick was fully prepared for his visit in 1891. He bought a pound of home-manufactured candles and also had a magnesium ribbon lamp, quite a complicated and bulky item of equipment, to lighten up the cave. The lamp featured a magnesium ribbon that was unwound on a clockwork movement. Once lit and with the clockwork movement running it would produce a continuous very bright light reflected by a mirror onto the area to be illuminated and, ideally, photographed. McCormick was not disappointed with what the lighting revealed: 'icicles sparkled with the brilliancy of ten thousand diamonds of the richest and purest water.' Douglas Hill Scott and the American W.S.C. Russell also recommended visiting similar lava features termed 'tintrons' around Mývatn and Húsavík. These are lava chimneys or 'hornitos', usually only a few metres high, and Russell talks of ascending one and looking 'deep, deep down' into its forbidding interior 'hung with lava stalactites'. Hornitos are formed where lava is ejected from a lava flow, producing a hollow, conical mound.

Surtshellir lies in an extensive area of lava tubes produced in the predominantly *pahoehoe* lava of the Hallmundarhraun lava field, part of the Western Volcanic Zone, produced in the early 900s CE. The name 'Surt-shellir', first mentioned in the twelfth-century *Landnámabók*, or Book of Set-tlements, was derived from Surtur, the Norse god of fire, who was believed to have been present at the creation of the world and would be responsible for its demise in a fiery conflagration. Applying such a name would suggest that the settlers in the area of Borgarfjörður were aware of the volcanic origins of

the caves and named them appropriately. The lava field is named after a troll,
Hallmundur, found in *Grettir's Saga* (*Grettis saga Ásmundarsonar*), who was
believed to live in a cave near Langjökull or Eiríksjökull from where the lava
originated. Lava caves in Hallmundarhraun were described as the homes of
outlaws, and recent excavations within the system have supported this view,
finding a range of deposits and features including hidden entrances, sleeping
enclaves and doors. The site also has a grisly reputation as the place where,
in 1236, Snorri Sturlusson's youngest son Óraekja was mutilated, albeit not
terribly successfully, under the orders of Sturla Sighvatsson, as recorded
in *Íslendinga saga*, part of the *Sturlunga saga* series. The joint scientific and
cultural significance of the area has been used to establish a Saga Geopark
covering the Borgarfjörður region, seeking to join Katla and Reykjanes as the
third UNESCO-recognized Geopark in Iceland.[7]

 Few, if any, geological texts make any reference to these early
observations of lava tubes; it is unfortunate that basic definitions usually
refer only to their presence on Hawaii, as if that was where they were
first discovered and described.[8] Acknowledgement is usually given to
the American Clarence Dutton (1841–1912), the first geologist to work in
Hawaii, who adopted the terms *pahoehoe* and *aa* in 1882.[9] Dutton was also
a founder member of the National Geographic Society in 1888 and would
no doubt have much enjoyed a visit to Iceland. He was a soldier and poet
as well as a geologist who described the Grand Canyon as bewildering and

  MOST UNIMAGINABLY STRANGE

overpowering, just the sort of polymath to have relished engagement with the Icelandic landscape. Since he was a *bon viveur* who loved cigars, drink and conversation, it is interesting to speculate on what a chance meeting with Lord Dufferin on the lava plains of Iceland might have produced. It is also likely that, as a self-confessed 'omnibiblical' person who devoured books, had he been aware of what had been written on the lavas of Iceland he would have given the observations of von Troil, Eggert Ólafsson and Bjarni Pálsson the platform they deserved.

By Mývatn the extensive labyrinth of Dimmuborgir (the Dark Castles or Dark Fortress) comprises an extensive rootless shield formed around 2,300 years ago when a lava lake 2 km (1¼ mi.) in diameter was formed, which subsequently drained – leaving a complex landscape of lava pillars. The lava lake was fed by magma travelling along 3 km (almost 2 mi.) of lava tubes from the site of the crater row, and the peculiar lava formations that have been left are thought to be the rims of old lava pools. The site was seen in folklore as connecting the earth with the infernal regions where Satan landed when expelled from heaven and is the home of the most brutal trolls. In particular, this refers to the half-troll, half-ogre Grýla, who has a taste for eating children, and her lazy husband Leppalúði, whose thirteen sons are the Yule Lads, or *Jólasveinar*. These mischievous figures visit on the days leading up to Christmas, carrying out pranks and leaving sweets or rotting potatoes as a gift for children, depending on how well they have behaved over the year.

Aa lavas are more viscous than *pahoehoe* and tend to flow in channels, often quite rapidly, up to 70 km (44 mi.) per hour (kph/mph), producing

Lava formations at Dimmuborgir.

ÞINGVELLIR, THE UBIQUITY OF BASALT AND THE MAKING OF ICELAND

Drekkingarhylur, Þingvellir.

a primordial ocean – hence the etymological homage to Neptune the Roman god of freshwater and the sea – within which rocks were formed by sedimentation. According to this theory granites were the oldest rocks but all rocks were formed in a similar way, including basalt. This view was opposed by the Plutonists, especially James Hutton, and later by Charles Lyell, who argued for their formation internally in a 'hot' Earth and a subsequent slow recycling of rocks between the land and the oceans. Underlying this principle was the idea of uniformitarianism, usually defined as 'the present is the key to the past', which proposes that the past could be inferred from observations of present processes, albeit requiring a much longer timescale, the possibility of almost inexhaustible 'deep time'. This exceeded the 5,800 years which had been adopted as the age of the Earth by the Irish cleric Archbishop Ussher, who in 1650 used the genealogy of the Bible as a chronological tool and proposed 22 October 4004 BC as the day of creation. Basalts, lavas, as exemplified at Þingvellir were therefore just as central to the debate as granites, and Iceland had plenty of the former, if lacking the latter.

Any traveller in the early and middle nineteenth century with geological interests would have been well aware of this conflict and Lord Dufferin was clearly of the Plutonist school. In this he was following the observations of von Troil in the previous century who, when describing what he found at Þingvellir, talked of cleft eruptions and subterranean fires. Von Troil also described in detail the basalt formations and the crystal structure of basalt columns with three to seven sides, having previously seen similar structures at Fingal's Cave on the Isle of Staffa in the Hebrides. One of the objectives of the expedition by Sir George Mackenzie in 1810, supported by the Royal Society of Edinburgh, which was no doubt keen to enhance the reputation of the key Scottish players in the debate, was to provide evidence in support of the Plutonist school of thought, and the cyclical theories of Earth history then being advocated. Accompanying Mackenzie, Henry Holland was known to be an enthusiastic mineralogist eager to play his part in demonstrating the veracity of Hutton's ideas. In an introduction to the Hakluyt Society edition of Holland's journal, Andrew Wawn points out Holland's 'burdensome' responsibilities to the scientific community of chronicling his Icelandic experiences but recognized 'the excitement of a young geologist let loose in the land of his mineralogical dreams'.

Henry (later Sir Henry) Holland made two visits to Iceland over sixty years apart but was another who was never really appreciated for his careful observations and insights. As a young and extremely enthusiastic geologist, although training as a doctor, he produced a very carefully written detailed journal of his 1810 travels with Sir George Mackenzie. Mackenzie's account, published in 1811 as *Travels in the Island of Iceland during the Summer of the Year 1810*, includes essays by Holland, but comparison between the published account and the manuscript of Holland's journal shows that Mackenzie relied heavily on the younger man's descriptions. Although Mackenzie was a mineralogist he tends to be remembered more for his descriptions of the Icelandic way of life and his considerations of the history and development, or rather lack of development, of this 'distinct and peculiar race of people', from saga times to the nineteenth century. Mackenzie's published work was very popular, quickly reprinted in 1812 and recommended by Pfeiffer, who stated that its 'sterling value . . . [was] appreciated everywhere'. It was revised in 1842, yet Holland still felt, quite fairly, that his contribution was not properly credited by his older colleague. When he returned to Iceland in 1871 at the age of 84, by then knighted and a physician to society in London, including Queen Victoria among his patients, he met William Morris in Reykjavík. Unfortunately the meeting was not a success, with Morris writing that the most interesting thing about Sir Henry was his age. Holland's full account was not printed until 1960, when an Icelandic translation was produced – probably the only travel account originally written in English that first appeared in print in Icelandic. It is therefore unfortunate that his observations and thoughts covering a wide range of subjects including education, literature, government and disease, within an account that engaged thoughtfully with the geological enigmas of the country, were rarely acknowledged by later writers.

Although Iceland was to play an important part in the geological debates of the nineteenth century, the role of Icelandic observers or chroniclers – for whom English in particular, but also German and French, were not the languages of publication – was overlooked. Charles Lyell quoted a description of the 1783–4 Laki eruption in his groundbreaking *Principles of Geology*, published in 1830,[15] but his source was Ebenezer Henderson's account of his visit in 1814–15, who in turn had seen a handwritten account of a visit to the site of the eruption produced by the Icelandic doctor and naturalist Sveinn Pálsson. Sveinn Pálsson's observations were not actually printed until 1945. The Danish government requested quite regular reports concerning the state of their colony and Sveinn Pálsson's observations came from such a report, but the handicap of writing in languages few could understand has meant that it was only in the later twentieth century that the findings of these Icelandic and Danish observers could be widely accessed.

Travellers other than Ida Pfeiffer writing about their visits to Iceland also spent many words outlining the variety of lava flows they found, especially their clear variations in age and the different structures they took on as they cooled at the surface. Travelling by horseback or on foot, they would have been acutely aware of such variations and, for geologists who had not had the opportunity to see the island, their descriptions provided valuable support for the Plutonists. Thus, as Charles Lyell's *Principles of Geology* became the more accepted basis for geological thinking, the volcanic landscape of Iceland provided an obvious focus for establishing a deeper understanding of Earth's history. Charles Darwin (1809–1882) had a copy of Lyell's book with him on the *Beagle* voyage, which he referred to extensively, and it is a loss for the earth sciences that Darwin did not visit Iceland before the voyage on which his interests started their transition from geology to biology.

Observations of Þingvellir and of lavas and basalts in general could be used to support Lyell's views, but it was still difficult for travellers and scientists to grasp the nature and immensity of deep time. Just how long was the geological timescale? Lord Dufferin summarized the problem: 'Ages ago – who shall say how long? – some vast commotion shook the foundations of the island.' Similarly, Pfeiffer, on crossing the extensive lava fields around Krýsuvík, commented: 'Who can tell whence these all-destroying masses of lava have poured forth, or how many hundred years they have lain in these petrified valleys?' It was not really until the twentieth century, and our understanding of radioactive isotopic decay, that ages of rocks could be determined with any scientific certainty and the real youth of the Icelandic landscape relative to the rest of the Earth could be revealed. Techniques such as K-Ar dating, which uses the regular radioactive decay of potassium into argon,[16] allowed direct dating of lavas over very long timescales, and the Icelandic basalts proved particularly valuable for this, accompanied by another geochronological approach based on palaeomagnetism.[17] As extruded basalts cool, they take on the alignment of the Earth's magnetic field, something that is constantly changing and has at times totally reversed, with the north and south poles switching. The extensive exposures of basalts in Iceland provided accessible rocks to collect and analyse for their palaeomagnetic signature, and showed that as you move away from the central part of the island, of which the active areas around Þingvellir form part, the palaeomagnetic records of the rocks provide mirror images going west and east, a feature also found in the rocks below the surface of the North Atlantic either side of the Mid-Atlantic ridge. We may know now that the ridge is a constructive plate margin with crust being formed and moving away in both directions, but it was the palaeomagnetic record that provided key evidence to support this idea.

Sequences of lavas were noted by both Henderson and Pfeiffer, who described numerous strata piled upon each other in cliffs from eastern and western Iceland. The geological potential of these lava piles that can be found on Iceland, especially in the eastern fjords, proved particularly attractive to a British geologist in the 1950s and 1960s. George P. L. Walker (1926–2005) spent every summer between 1955 and 1966 carrying out fieldwork in true British fashion, relying on a small tent and a van, often alone, and occasionally having to be rescued by accommodating and sympathetic farmers while crossing rivers. His meticulous and detailed mapping of the individual lava units allowed an interpretation of the build-up of the island by spreading from a central axis, akin to the later sea-floor spreading confirmed by the palaeomagnetic record in the early 1970s. Eastern Iceland proved an excellent if challenging location for such innovative research, having an estimated nine hundred lava flows extruded between 13.5 and 2 Ma – a thickness of 8.8 km (5½ mi.) with 10 m (33 ft) being added every 12–14 ka.[18] Walker's work was recognized in Iceland by his election to the Award of the Falcon (Knight's Class) in 1980, unusual for a non-Icelander, and by the housing of his notebooks, maps, photographs and other associated material in the Breiðdalssetur Research and Heritage Institute, otherwise known as the Walker Institute, in Breiðdalsvík, eastern Iceland[19] – one of the more idiosyncratic exhibitions in a country that also boasts phallological, sorcery and witchcraft, shark and whale museums.

Columnar basalts, or *stuðlaberg*, have proved an important visual image and intriguing natural element of the Icelandic landscape. *Kirkjugólf*, or Church Floor, the small (80 m²/860 ft²) area of columnar basalt pavement just to the east of Kirkjubæjarklaustur, was visited by Henderson, who commented on the remarkable regularity of the pentagonal surface. He had already seen basalt cliffs at Rauðaberg further east, called *Tröllahlað* (Giant's Wall) by his guide, which impressed him: 'I almost fancied myself amid the ruins of some of the noblest structures of ancient Grecian architecture. The pillars were piled one above another with the most perfect exactness.' As he saw different areas of columnar basalt he noted their geometrical variations, varying from pentagonal to octagonal, and was inspired in this case to invoke Psalm 104: 'O Lord, how manifold are thy works.' The columnar structure originates as the basalt cools and contracts. If the contraction occurs regularly then a hexagonal pattern arises; slight irregularities lead to a different number of faces. Where water or ice is involved with the basalt on cooling, a more irregular series of columns is produced, often found overlying the regular colonnade and termed an entablature, or *kubbaberg* in Icelandic. The stunning basalt cliffs at Stuðlagil in Jökuldalur, eastern Iceland, have only recently become an attraction given their relative inaccessibility, and also due to their increased exposure in 2015 with the falling level of the Jökulsá á Dal, following the building of the Kárahnjúkavirkjun power plant. For earlier travellers with a yacht, Stappi (Arnastapi) and the

Svartifoss waterfall, Skaftafell, a popular tourist site exhibiting well-developed columnar basalt.

Gatlekkur coastal arch, part of the Lóndrangar cliffs around the Snæfellsnes peninsula, provided the most readily accessible view of excellent basalt columns. Sailing around the area in 1834 John Barrow was able to state: 'here the theory of the Neptunists . . . falls to the ground', echoing the view of Charles Lyell on the Neptunists, that 'it would have been impossible for human ingenuity to invent one [theory] more distant from the truth.' The variety of forms of basalt and their picturesque locations in Iceland proved attractive both visually and intellectually, stimulating intellectual curiosity or confirming the existence of a divine Creator responsible for 'the more secret recesses of the great cabinet of nature'.

Apart from basalt, one of the commonest extrusive rocks found in lava flows is rhyolite, which provides a colourful alternative to the unending grey. Rhyolite is generally much lighter in colour but can vary from grey and yellow to pink and red and is defined by its higher silica content of over 67 per cent, compared to less than 52 per cent for basalt. Some of the most impressive Icelandic landscapes in terms of colour and form have significant rhyolitic components, as at Landmannalaugar, the largest area of rhyolite in the country. The German chemist Robert Bunsen (1811–1899) noted that rhyolite and basalt could be found together and used that observation to argue for different magma sources, contributing to the debate concerning the origins of the various lavas found at the surface and what they could mean for Earth history in volcanic regions – a precursor of the hotspot debate of the twentieth century. Although initially presented by him in 1851, his ideas on rhyolite formation took over a century to gain widespread acceptance in the geological community.[20] The most extreme glassy form of rhyolite produced by rapid cooling is obsidian; Henderson was delighted to see Hrafntinnuhryggur near Krafla, a mountain composed mainly of obsidian, or *hrafntinna*, where he managed to collect specimens, some of which he had to leave behind as they were too large to carry. This mountain was well known and often noted as one of the wonders of the country by other travellers who made it to northern Iceland.[21]

Magma travels through the crust in conduits known as dikes and as it cools these solidify, forming intrusive rocks. Dikes occur in swarms, which intensify with depth, and the eroded cliffs of the fjords in eastern and western Iceland reveal impressive dike forms, usually comprising basalt. Such swarms can be over 50 km (30 mi.) long and 5–15 km (3–9 mi.) wide and eventually produce lava flows after they breach the surface. A further intrusive rock, gabbro, which has the same composition as basalt but provides a visually different almost granite-like appearance, can be found at a number of locations, such as the intrusion that forms Hvalsnesfjall, part of an exhumed Tertiary volcano in the Lón area of southeast Iceland. Gabbro

MOST UNIMAGINABLY STRANGE

Series of vertical dikes in a cliff face on the northern side of the Snæfellsnes peninsula.

now offers one of the most bizarre tourist souvenirs to be bought in the country. Nine small blocks of Icelandic gabbro can be bought for around $45 to be placed in a freezer and then used to chill drinks – 'on the rocks'!

Current geological understanding suggests that Iceland first appeared around 24 Ma, in the Miocene Epoch of the Tertiary period,[22] and has been extending at a rate of around 2 cm (¾ in.) a year, 1 cm (½ in.) in each direction. The spreading has often occurred as complex rifting events, not a gradual, continuous movement. It is interesting to note that over the period of the Krafla Fires in northern Iceland between December 1975 and September 1984 – a basaltic fissure eruption that occurred as a series of 21 rifting phases – the overall total widening was 900 cm (350 in.), an annual rate equivalent to the long-term rate of spreading.[23] The origins of Iceland lie further back in time, with the development of the Ægir ridge and the gradual separation of the plates between Greenland and Scandinavia around 70–60 Ma. The oldest exposed rocks in Iceland are thought to be around 16–14 Ma and over nine hundred distinct lava flows have been identified. Strata of different origins occur between the flows. In the earliest of these, fossiliferous sediments can be found representing humid temperate forests of North American, eastern Asian and western Eurasian origin, with species no longer found in Iceland such as sequoia, *Fagus* (beech), sassafras, magnolia and rhododendron. Gradual cooling can be seen as the island of Iceland grew, and after 12 Ma plant migration appears to have been limited to European sources, although with the closing of the Panama isthmus between the Atlantic and Pacific Oceans around 3.6 Ma, Pacific marine molluscs arrived around Iceland via the Bering Strait and the Arctic Ocean.[24] The lava sequences also include the earliest direct evidence for glaciation in Iceland at around 4 Ma, prior to the start of the Quaternary period (Pleistocene) at 2.6 Ma.[25] This latest geological period has been characterized by the cyclical growth and decay of ice sheets, although it is likely that ice first appeared at high altitudes on Iceland at least as early as 7 Ma. The evidence for glaciation takes the form of tillites, lithified glacial tills, or other glacigenic deposits, and correlation of these layers across the country suggests the occurrence of at least twenty different glacial cycles (see Chapter Eight).[26]

Fossils from the Miocene beds had been found by both locals and visitors over the centuries, especially lignite – fossilized organic remains that could be used for fuel. Lignite, or *surtarbrandur*, another reference to Surtur as the God of Fire, has actually been mined at seven locations around Iceland, especially in the northwest, where a mine was operated at Sýðridalur, near Bolungarvík, between 1917 and 1921, taking advantage of the demand for coal during the First World War. As fuel, the Icelandic lignite is relatively poor given its low carbon content of around 60 per cent, due to the amount of volcanic ash mixed in with the plants.[27] Uno von Troil was aware of the lignites, and their stratigraphic position caused some later discussion of how their presence could be accommodated within the biblical timeframe for the age of the Earth. John Barrow reports on von Troil finding lignite

Miocene plant fossils of maple (*Acer crenatifolium*), hazel (*Corylus* sp.) and magnolia (*Magnolia* cf. *reticulata*) collected by Jóhannes Áskelsson from Brjánslækur, northwest Iceland.

MOST UNIMAGINABLY STRANGE

under several strata of solid lava and entering into a discussion with the Scot Patrick Brydone about the implications this would have for their age. Brydone had ascended Mount Etna in 1770 and observed seven distinct lavas, on each of which he calculated it would take 2,000 years for soil to form, making the oldest lava at least 14,000 years old – well beyond the suggested age of the Earth. By implication, the ages of the lignite in Iceland would similarly be beyond their biblical age estimate. Von Troil counselled Brydone not 'to pretend to be a better historian than Moses', but by way of retort the Scot felt that his conscience would not let him believe Etna to be so young and that he must 'stick to his theory at the expense of his canonicals'.

In the 1750s Eggert Ólafsson identified fossil leaves of birch, rowan, willow and possibly oak in lignites from the eponymous Surtarbrandsgil in northwest Iceland, and in the nineteenth century, as part of a wider survey of potentially utilizable minerals and deposits, Jan Steenstrup (1813–1897) collected a number of fossils and sent them for curation to the Geological Museum in Copenhagen. Steenstrup was the first person to make the important observation that the early fossils came from species with American affinities, unlike the plants now found in Iceland that are almost all of European origin. Steenstrup's collection was supplemented with samples collected by the German Gustav Georg Winkler, the 'Ultra Neptunist', who was sent to Iceland in 1858 by Maximilian II of Bavaria, a monarch renowned for his wide interests in the arts and natural sciences. Finds of some of the earliest fossils were later made by two Icelandic geologists, Guðmundur Bárðarsson (1880–1933) and Jóhannes Áskelsson (1902–1961), both of whom identified beech; Áskelsson's from some of the oldest known sites in Selárdalur.[28]

Apart from the fossils Iceland was also known as a source for a range of minerals. Iceland spar (Iceland crystal or *silfuberg*), a clear calcite with regular crystals, was originally considered unique to Iceland, hence the name, and was valued for its use in microscopes due to its double-refractive properties. Iceland spar was mined as early as the seventeenth century at Helgustaðir. It has also been suggested that it may have been used for navigation by the earliest settlers, as it could be used to estimate the position of the Sun in cloudy conditions, hence its Icelandic name of *sólarsteinn*. The mineral was crucial to German optical science in the nineteenth and early twentieth century, and was known as *isländischer Doppelspat(h)*. Iceland spar was acquired by foreign visitors and seamen such as French fishermen, and it was sporadically mined commercially, especially at a location at Hoffell in southern Iceland in 1910. Mining had ceased by 1925 owing to falling demand, possibly due to stockpiling by users.[29] Over three hundred different minerals have been found in Iceland

and there are two small mineral collections on public view from private collections at Hveragerði in the south and Petra's Stone at Stöðvarfjörður in east Iceland, the latter built up by Ljósbjörg Petra Maria Sveinsdóttir from over eighty years of collecting.[30] Born in 1912, she had collected stones from her childhood, but had nowhere to keep them together. Following her marriage in 1945, on the same day as both her sister and brother married, she and her husband Nenni started to build a family house, which became the location for all of her collection. On Nenni's early death at 52 in 1972, Ljósbjörg Petra Maria Sveinsdóttir decided to open her house and the collection to anyone who wanted to visit, and this has been the position to the present. In 2003 over 20,000 visitors passed through what was, and still is, a private house. Petra died in 2012, just under a year short of her 90th birthday, having seen the publication of her book about her life in Icelandic, and leaving a worthy addition to the diversity of Icelandic museums.

While in the northwest of Iceland, Ebenezer Henderson came across an area from which particular minerals had been collected, which he said was by then, the early nineteenth century, already well known in Europe. Although he considered the source inexhaustible, he observed that 'the greatest waste had been made' in extracting the rocks. The minerals in question were secondary minerals, or zeolites, formed in cavities within the basalts as they cooled and lost their water content. The remaining water dissolved

MOST UNIMAGINABLY STRANGE

the minerals, which then cooled slowly and produced crystals of varying size and colour. George Walker was able to show that the position of zeolites relative to an extrusive event was determined by depth and temperature; if temperatures were too high then they would not form. This enabled him to predict the original surface of the lava pile under examination.

Quite why the island of Iceland has formed where it is, and in its present geological form, is still not entirely agreed between geologists. Following the acceptance of plate tectonics as the fundamental explanation for the Earth's geological structure, there has been some debate over the proposed existence and nature of discrete areas of high volcanic activity or 'hotspots'. It has been argued that Iceland represents the juxtaposition of a plate margin with a hotspot fed by a mantle plume, leading to the intense and continuous production of magma and its appearance at the surface where it cools. Over the last 20+ Ma this has produced the Iceland Basalt Plateau, covering 350,000 km^2 (135,000 mi.^{2}) and rising up to over 2,000 m (6,500 ft) above the sea floor, of which 30 per cent or 103,000 km^2 (40,000 mi.^{2}) is currently above sea level. The existence of mantle plumes is not supported by all geologists. Seismic techniques based on tomography, which utilize the various seismic waves produced by earthquakes to image the structure of the subsurface, have not provided unambiguous evidence for the presence of such a plume under Iceland. The specific location of the Iceland hotspot is also uncertain. The original view was that it was located under Vatnajökull, propagating southwest to Vestmannæyjar (the Westman Islands). An alternative view is that the hotspot lies away from Iceland, and feeds magma towards Iceland by lateral movement at the boundary between the lithosphere and the highly viscous asthenosphere, in the upper part of the mantle. A further view is that there is no hotspot or mantle plume at all, but that Iceland is an area of high mantle fertility associated with an ancient subduction zone, which is crossed by a spreading ridge. This subduction has been responsible for the remnants of an old oceanic plate surviving in the uppermost mantle, with enhanced melting causing the concentration of volcanic activity.[31]

Testing these theories relies on geophysics reconstructing the properties of various layers below Iceland and observing the geochemistry of the extruded lavas to determine their likely origins. Traces of rocks from different sources deep below the surface can be found as xenoliths, alien rock found within the extruded magma. Where such rocks cannot have been produced within the magma, it is necessary to provide an alternative source, either country rock through which the magma passed, or continental/oceanic crust incorporated within the mantle. Sampling of extruded material from the 1963–7 Surtsey eruption showed the presence of complex xenoliths, derived from ice-rafted debris carried from Greenland and deposited on the sea floor over the source of the volcano, including at least one rock derived from ship's ballast dumped in the same location. Ship's ballast was also responsible for Icelandic rocks finding their way to Europe. Lava was used as ballast for the *Resolution* in 1772 and when it returned to England Sir Joseph Banks used this for the moss garden at Kew and the rockery at the Apothecaries' Garden at Chelsea, now known as the Chelsea Physic Garden.

Mantle plumes and palaeomagnetism seem a far cry from the observations of von Troil, Mackenzie, Dufferin, Pfeiffer and others, which shows just how far geology has progressed over the last century or so, but to read their accounts with the benefit of current understanding is to recognize just how observant and deep thinking they were. They were visiting a country that was certainly alien to most of their previous experiences but was also remarkably uniform in character – lava was everywhere – and it was not always particularly scenic. In studying the journals of William Morris, Andrew Wawn observed that he used the word 'grey' over a hundred times with a wide range of descriptors (very dark grey, dreadful grey, greyer than grey, inky grey and so on), although he was not using the colour to describe only the land, but the sea and sky.[32] Morris's use of 'grey' upset his friend the poet and fighter for Icelandic independence Jón Sigurðsson, to whom he sent two poems to translate for publication after returning home, as the Icelander considered that the poems presented a rather dismal and gloomy view of Iceland, whereas for Morris, very much an *Íslandsvinur*, a Friend of Iceland, greys were not at all dull and often formed an important backdrop to brighter primary colours in his designs.

As an artist Morris might have been expected to have a high sensitivity to colour; however, it is not just the tones but the textures of the lava that were important. The artist Jóhannes Sveinsson Kjarval was noted for his idiosyncratic depictions of the shapes and forms of the lavas in his paintings of the Icelandic landscape. Like Walker, he spent a lot of time out in the landscape using his knowledge of the microform of lava fields to provide shelter, recognizing the diversity of structures taken on by the cooling lavas. Reflecting the morphology of the landscape as he saw it became an important part of his work. Just as those walking or riding across endless lava plains were aware of relatively subtle variations in the form of the surface, so too did later scientists begin to understand the significance of these complex variations in terms of the processes that gave rise to the variety of lava flows, as they reflected changes in the nature of the original magma and its sources as well as differences in rates of extrusion and subsequent cooling. Beyond the scientific questions, lava fields were also seen to have a political dimension, relating to the Icelandic independence movement that began to develop during the nineteenth century, and in which Jón Sigurðsson played such an important part. To Sigurðsson and other nationalists, the barren nature of the lava fields was seen as representative of the stultifying effects of Danish rule, particularly manifest in the trade monopoly that kept out influences from other European and American sources. The restrictions of the monopoly were apparent in the aftermath of major environmental disasters, such as the eruptions of Laki in 1783–4 and Hekla in 1845. Glaciers and volcanoes, however, were symbols of Icelandic independence. In 1902 the academic Jón Stefánsson (1862–1952), not to be confused with the living author Jón Kalman Stefánsson, when speaking to the Philosophy Society of Great Britain, firmly laid the blame for Iceland's troubles in the eighteenth and nineteenth centuries at the door of the Danish colonial government in combination with environmental catastrophes: 'Nature seemed in league with man to render Iceland uninhabitable.'[33] Jón Stefánsson's view was that Iceland was geologically and culturally aligned with the British Isles and that if only they could throw off the burden of Danish rule all would return to greater harmony between man and nature in the country, something akin to conditions at the time of first settlement. Jón Stefánsson is also famous for describing Iceland in the same paper as a 'living Pompeii where the northern races can read their past', the first time such an analogy was used, although as we shall see this was not the only time that comparisons have been made between Iceland and Pompeii.

The landscape of southern Iceland has been modified over very recent geological history by a number of extremely large lava flows. The Þjórsá lava, the oldest of a series of eight flows from Veiðivötn originating north of Landmannalaugar and which dates back to around 8000 BP,[34] is the most voluminous in Iceland, comprising 21 km³ (5 mi.³) of lava stretching over 140 km (87 mi.), and is believed to be the largest Holocene lava field in the world. The Eldgjá lava, which was 'discovered' by Þorvaldur Thoroddsen in 1883, originated from what in 934–40 CE was a 75-kilometre-long (46½ mi.) fissure from the Katla system, and proved exceptionally productive, 14 km³ (3½ mi.³) of lava being extruded in a series of up to twenty episodes. The Reykjanes peninsula is characterized by several historical lava flows that can be identified in documentary records. The Ögmundarhraun flow is traditionally thought to have been extruded as part of the Krýsuvík Fires, between 1151 and 1158; one of the few Icelandic lava fields named after a person, its age has recently been questioned using palaeomagnetic dating, which provides a cautionary tale when linking natural events with documentary sources. Ögmundur was believed to be a 'berserker' who excavated a road across the lava field in order to convince the local farmer that he should marry his daughter. In Old Norse, berserkers were warriors who fought with what has been described as a trance-like fury, forsaking any form of armour. The farmer regrettably was still not convinced of Ögmundur's suitability for his daughter and killed and buried him when he was tired after completing the road. A similar story is used on the Snæfellsnes peninsula to name the Berserkjahraun, or Berserker lava field, which is known to have been produced 3,000–4,000 years ago, well before the time of the Vikings.

Lava flows dominate the landscape of Iceland either in the form of the extensive lava plains or in impressive cliffs eroded by the sea or by ice. Where the sequences are not so thick the landscape tends to be more subdued and is punctuated by different volcanic forms from shield volcanoes to the almost ethereal structures of palagonite formed by sub-glacial eruptions, but predominantly Iceland is a very 'horizontal' country: the strata have not been folded and distorted as in large areas of Europe – what the poet Alyson Hallett has described as 'outcrops of horizontalled time'[35] and W. H. Auden saw as 'having a curious shape like vaulting-horses in a gymnasium'.[36] For a country in which, according to the Elf School in Reykjavík that offers classes in all matters related to elves and hidden folk, 54 per cent of the population believe in *huldufólk* (hidden folk) such as elves, trolls and a variety of other beings, each new lava flow can be seen to create

Elf houses near Strandakirkja.

MOST UNIMAGINABLY STRANGE

new homes for them with different-coloured stones providing the houses;
for instance red for enthusiastic and energetic elves and yellow for intelligent, optimistic and playful people. In Hafnarfjörður in the southwest,
which claims to have Iceland's largest settlement of *huldufólk*, there is an elf
garden in a lava park and it is possible to go on a Hidden Worlds Tour.[37] The
link between lava and *huldufólk* is very strong and reinforces the essence of
the variety of lava at the local scale as distinct from its apparent greyness
and dullness when seen as a large lava plain covering the landscape.[38]

The physical structure and form of the country is not only found
striking by visitors but is a feature that has cultural resonance. Hallgríms-
kirkja, the church that dominates the Reykjavík skyline, was designed to
reflect the forms of the Icelandic landscape, with buttresses resembling
basalt pillars. The startling design of the Harpa concert hall in Reykjavík
consciously reflects the crystal structure of basalt. Ólafur Elíasson
designed the facade using quasi-bricks, twelve-sided crystalline panels,
which reflect and consume light as well as mimicking the geology. As the
architects Henning Larsen explain:

> the halls form a mountain-like massif that, similar to basalt rock on
> the coast, forms a stark contrast to the expressive and open facade.
> At the core of the rock, the largest hall of the Centre, the Concert
> Hall, reveals its interior as a red-hot centre of force.[39]

William Morris would recognize and echo these sentiments.

HEKLA AND HELL: THIRTY VOLCANIC SYSTEMS AND COUNTING

'HEKLA PERPETUIS
DAMNATA ESTIB. ET NI
UIB. HORRENDO BOATU
LAPIDES EVOMIT'

'THE HEKLA, PERPETUALLY CONDEMNED TO STORMS AND SNOW, VOMITS
STONES UNDER TERRIBLE NOISE.'[1]

There is no record of anyone climbing the volcano Hekla before 1750, perhaps because of its reputation. On the map produced by the Flemish cartographer Abraham Ortelius in 1585 and published first in 1590, it is the only volcano represented and shown in an extremely fiery form with the inscription: '*Hekla perpetuis / damnata estib. et ni / uib. horrendo boatu / lapides evomit*' (The Hekla, perpetually condemned to storms and snow, vomits stones under terrible noise). The map was almost certainly drawn originally by Guðbrandur Þorláksson, the Icelandic mathematician and Bishop of Hólar in northern Iceland, who had lived at Skálholt near Hekla and must have had a good knowledge of this area of Iceland. A belief in Hekla as the gate to hell can be traced back at least to the twelfth century, when Herbert of Clairvaux wrote in *Liber der miraculis* that in comparison to the fire-kettle of Mount Etna, which 'is called the vent of Hell and to which, as has often been proved, the souls of the dying, condemned to burn, are daily dragged', Hekla was 'this immense pit of Hell'. In 1536 the German itinerant scholar Jakob Ziegler (1470–1549) described it as a fissure always ablaze with spirits trapped in the fire and with other spirits walking around, a description put more eloquently two decades later by the German scholar and physician Caspar Peucer (1525–1602), who wrote in 1554:

> Out of the bottomless abyss of Hekla Fell, or rather out of Hell itself, rise miserable cries and loud wailings, so that these lamentations can be heard for many miles around. Coal-black ravens and vultures

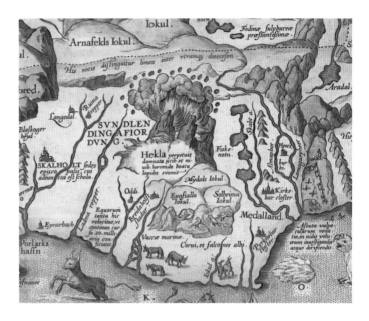

hover around this mountain and have their nests there. There is to
be found the Gate of Hell . . . there can be heard in the mountain
fearful howlings, weeping and gnashing of teeth.[2]

Strong stuff indeed and likely to put off any true believers from investigat-
ing too closely.

Nevertheless, in 1593 Arngrímur Jónsson in *Brevis commentarius de
Islandia*, his *Defence of Iceland*, argued that Hekla was volcanic rather than
diabolic and tried to reassure visitors and his fellow countrymen and
women that the noises coming from the mountain were not the groans of
lost souls being dragged into hell.[3] Those first to ascend the mountain in
the eighteenth century, Eggert Ólafsson and Bjarni Pálsson on 20 June 1750,
were also keen to dispel the myth, so by the time travellers more commonly
included an ascent of the volcano as part of their itinerary they described
their experiences with science in mind, rather than the possibility of
eternal damnation. In case visitors were fearful of experiencing physical
danger from eruptions, Niels Horrebow in the same year wrote: 'An
eruption very seldom happens, and even when it does, it occupies but
a small tract of time. Travellers cannot therefore be much obstructed by
it.' Although Sir Joseph Banks is principally remembered as a botanist,
his passport for the visit in 1772 granted by the Danish envoy in London

Unknown artist, *Guðbrandur Þorláksson, Bishop of Hólar*, 1620, oil on canvas.

was to allow him to observe Mount Hekla. Prior to arriving in Iceland, he made enquiries about how best to view the mountain and what else was 'burning'. He set up his base in Iceland at Hafnarfjörður but found there was no 'burning' taking place at the time and decided to make what he believed to be the first ascent of the volcano. In far from clement weather and covered in ice they made the summit and found 'scenes of ashes and desolation all round almost inconceivable'.

Almost a century after the first ascent, Ida Pfeiffer made the summit with a guide, after a 'fearful journey', partly on horseback although she found walking preferable. With her experience of climbing Vesuvius, she immediately set off looking for a crater, but found only a number of clefts and fissures below the summit. Undaunted, she still recognized lavas marking different earlier eruptions, noting that Hekla had 'the blackest lava and the blackest sand', and encapsulated her feelings about the volcanic landscape surrounding her in an impressive descriptive passage neatly summing up the desolation and beauty as she saw it: 'Here, from the top of Mount Hecla, I could see far into the uninhabited country, the picture of a petrified creation, dead and motionless, and yet magnificent.' In her description of the summit she incurred the wrath of Pliny Miles, who queried whether she actually got there because he found a crater on his ascent in 1853. In many ways Pfeiffer's account is more like the later one of the Reverend Frederick Metcalfe, although he does mention the presence

MOST UNIMAGINABLY STRANGE

Photograph of Ida Pfeiffer taken by Franz Hanfstaengl around 1856, shortly before she died. The globe recognizes her importance as a traveller of international renown.

of a crater. Miles obviously did not like Pfeiffer and called her 'the old Austrian dame, that Madame Trollop'. More seriously he questioned the truthfulness of other aspects of her travels,

> where she does not knowingly tell direct falsehoods, the guesses she makes about those regions she does not visit – while stating that she does – show her to be bad at guess-work, and poorly informed about the country.

Miles's geological understanding was certainly poorer than most of his contemporaries – for instance, he favoured a single submarine eruption to account for Iceland – but he was generally considered a truthful correspondent although he had a 'nervous-sanguine temperament'.[4] It does, though, seem likely that Miles was not averse to talking up the dangers of his visit to Iceland for his readership, and nowhere was this more apparent than on Mount Hekla with its 'perpendicular slopes and burning craters', where even the accompanying dog trembled with fear. Pfeiffer's account of her 'fearful journey' suggests a more straightforward, albeit uncomfortable experience, as she often fell and cut herself on the lava, certainly not anything seriously life-threatening.

The desolate interior of Iceland seen by Pfeiffer from Hekla's summit had proved of interest to Icelanders almost since settlement times,

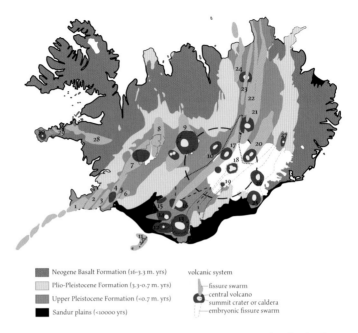

Volcanic systems in Iceland. The volcanic zones are named in the text. Individual systems are numbered: 1. Reykjanes/ Svartsengi; 2. Krýsuvík; 3. Brennisteinsfjöll; 4. Hengill; 5. Hrómundartindur; 6. Grímsnes; 7. Geysir; 8. Prestahnjúkur; 9. Hveravellir; 10. Hofsjökull; 11. Tungnafellsjökull; 12. Vestmannæyjar; 13. Eyjafjallajökull; 14. Katla; 15. Tindfjöll; 16. Hekla/ Vatnafjöll; 17. Torfajökull; 18. Bárðarbunga/Veiðivötn; 19. Grímsvötn; 20. Kverkfjöll; 21. Askja; 22. Fremrinámur; 23. Krafla; 24. Þeistareykir; 25. Öræfajökull; 26. Esjufjöll; 27. Snæfell; 28. Ljósufjöll; 29. Helgrindur (Lýsuskarð); 30. Snæfellsjökull. From Thorvaldur Thordarson and Ármann Höskuldsson, *Jökull*, volume LVIII (2008), which was based on Jóhannesson and Sæmundsson's 1998 *Geological Map of Iceland*. 1:500,000. *Bedrock Geology*.

Neogene Basalt Formation (16-3.3 m. yrs)

Plio-Pleistocene Formation (3.3-0.7 m. yrs)

Upper Pleistocene Formation (<0.7 m. yrs)

Sandur plains (<10000 yrs)

volcanic system

fissure swarm

central volcano

summit crater or caldera

embryonic fissure swarm

North Volcanic Zone (NVZ), East Volcanic Zone (EVZ), Mid-Iceland Belt (MIB), Öræfajökull Volcanic Belt (ÖVB) and the Snæfellsnes Volcanic Belt (SVB).[10] Over the last 1,100 years since settlement, 80 per cent of eruptions have occurred in the EVZ, producing an estimated 87 km³ (21 mi.³) of Dense Rock Equivalent (DRE), double the rate of Hawaii, with 71 km³ (17 mi.³) deriving from the Katla, Hekla and Grímsvötn systems. Overall during the Holocene, or the last 11,700 years, there have been an estimated 2,400 eruptions producing 566±100 km³ (136±24 mi.³) of erupted magma.[11]

As the third most active volcano in Iceland after Grímsvötn and Katla, Hekla is unusual in having repeated eruptions from the same fissure and despite having a typical stratovolcano shape, when seen from one direction, it is often described as boat-shaped. Henderson was not impressed with the mountain as a landscape feature:

> There is little in the appearance of Hekla to attract the notice of the traveller, nor do I conceive there is anything about Hekla that is calculated to make an indelible impression on the memory, except an actual eruption.

Neither was Sir Richard Burton, who called it 'a commonplace heap'.[12] Nevertheless, depictions of Hekla sometimes paid little attention to the actual form; it was what Hekla represented that counted and a little artistic licence did not go amiss. A case in point was a large tableau constructed at the Surrey Zoological Gardens in London in 1839, shortly after John Barrow's account was published. The gardens had been laid out in 1831 and featured the largest conservatory building of its time. Associated with this were displays on large 24-metre-high (80 ft) painted backdrops. The 1839 representation of Iceland and Mount Hecla [*sic*] followed one of Vesuvius from 1837 and featured a highly stylized painting of a smoking Hekla with a foreground of icebergs. During the evening this mural was supplemented by fireworks, principally rockets and Roman candles, which were extremely popular and displayed on a daily basis.[13]

Since settlement there have been eighteen reported eruptions of Hekla producing 10 km³ (2½ mi.³) of lava, with evidence for at least thirty in the last 7,500 years of the Holocene. The magma chamber is around 3 km (2 mi.) deep and eruptions occur in one of three forms. They can be highly explosive Plinian eruptions,[14] which produce large volumes of tephra (see below for a definition of tephra), such as Hekla 4 (4,300–4,200 BP) and Hekla 3 (3,100–3,000 BP). Ash from the older Hekla 4 eruption covered over 80 per cent of the island. There can also be a mix of explosive and effusive eruptions as in 1947, which was commemorated in a series of postage stamps the following year; or eruptions that are almost entirely effusive and that have little effect beyond the summit area, as in February–March 2000. Most eruptions are small and characterized by a brief initial explosive phase, which may only last a few hours, followed by a few days of fissure eruption. This behaviour is probably why Horrebow was keen to reassure travellers of the infrequency and brevity of eruptions, as Hekla would have been the volcano most likely to be seen by any visitors arriving at Reykjavík and having little time to explore the country as a whole.

The 1104 eruption of Hekla was well attested by settlers who had been in the country for over two centuries, as five farms were lost due to the tephra fall at distances of up to 70 km (44 mi.) away from the mountain. There is also a very good account of the 1300 eruption written in the fourteenth-century *Lögmannsannál* by Einar Hafliðason, an administrator of the episcopal see at Hólar, who describes the

> coming up of fire in Mount Hekla with such violence that the mountain split, so that it will be visible as long as Iceland is inhabited . . . the wind . . . carried such sand, that no-one inside or outside could tell whether it was night or day.

Hekla 4 tephra deposit at
Trjáviðarlækur, 15 km ssw
of Mount Hekla.

The sand noted by Einar Hafliðason was volcanic ash produced largely
during the initial explosive eruptive phase and even at this time in such
records Icelanders would distinguish between pumice (*vikr*), sand (*sandr*)
and ash (*aska*). These are now known as tephra (from the Greek for ash,
τεφρα, first recognized by Aristotle in the 4th century BCE on the Lipari
Islands off Sicily), thanks to the pioneering work of Sigurður Þórarinsson,
Iceland's 'best known geologist' and the most widely known twentieth-
century Icelandic earth scientist, who was born in Vopnafjörður in 1912 and
died at the relatively young age of 71 in 1983. His links with tephra go back
to his father, whose imminent birth delayed the abandonment of the family
farm by Sigurður Þórarinsson's grandparents in the face of the 1875 eruption
of Askja.[15] Originally largely self-taught, being brought up in a parsonage
until his father died and he and his siblings were dispersed to local farms,
Sigurður Þórarinsson owed much to the overseeing of his education by a
local clergyman who checked on his progress and ensured that he covered
sufficient topics to gain a place at a high school in Akureyri, to which he
departed at the age of fifteen. As the best student in his year he was awarded
a stipend to study in Copenhagen but after a year moved to Stockholm where
he stayed for thirteen years, only returning to Iceland in the summers to
carry out fieldwork, as when he accompanied Hans Wilhelmsson Ahlmann
in 1936. He returned to Iceland in 1944 via Britain and a transatlantic convoy.
Though he initially wrote his thesis in Swedish in 1944, much of his work was
produced in English – ensuring, unlike his predecessors, that he was given
due widespread academic acknowledgement for his innovative work.

Sigurður Þórarinsson utilized the unique nature of each ash layer
produced by eruptions as a chronological tool to examine the spread of
former eruptive plumes, particularly concentrating on Hekla. By this
method, known as tephrochronology, he was able to reconstruct and
date eruption sequences for different Icelandic volcanoes, and laid the
foundations for what is now an essential and universal chronological tool
in the earth sciences. In the true tradition of Icelandic science, Sigurður
Þórarinsson was something of a polymath, being well known for his
popular songs and not just for his breadth of geological interests. He was
very much a public scientist, passionate about conservation, but also
unassuming. The glaciologist Helgi Björnsson, in an obituary of the Grand
Old Man of geoscience in Iceland, commented that he appeared unaware
of his eminence and significantly that he was 'an artist at heart, both as
scientist and poet', a man with a 'sublime imagination'.[16] It will come as
no surprise to learn that the title of one of his academic papers about the
long-distance transport of tephra to Scandinavia begins with 'Greetings
from Iceland'.

Over recent years tephrochronology has become one of the most powerful dating tools available in studies of recent climatic and environmental change, particularly over the most recent geological period, the Quaternary (the Pleistocene Epoch), the last 2.58 Ma, which is characterized by the cyclical growth and decay of great ice sheets. Layers of tephra are found not just in geological sequences close (proximal) to the source volcano, but as isolated microscopic volcanic shards, cryptotephra, in sediments from the bottom of the ocean and lakes, in peat bogs and crucially in ice cores at distant (or distal) locations. Ice sheets and ice caps thicken by the incremental addition of snow, which slowly turns to ice, trapping the ash as a discrete horizon. The layers of ice have various other signals within them based on the chemical signature of the snow that forms the ice, enabling the identification of annual and even seasonal layers. Even without tephra shards, peaks of acidity in the ice can reflect eruptions rich in sulphur and be linked to potential source eruptions. The power of this technique in association with other chronological tools such as tree-ring dating, or dendrochronology, has been demonstrated at many locations but recent Icelandic examples serve to illustrate its potential.

In 2003 a flood of the river Þverá in southern Iceland revealed what is known as the Drumbabót forest, stumps of birch trees up to 60 cm (24 in.) high formerly growing on the site which had been killed in a previous *jökulhlaup*, or glacier flood, from Katla, the volcano that lies beneath the Mýrdalsjökull ice cap, 50 km (31 mi.) east of Mount Hekla. The trees had their bark intact and discernible tree-ring records, showing that they died during the winter. From one particular 76-year-old tree, it was possible to identify an annual ring with a clear cosmogenic spike, an anomalous peak in radiocarbon produced in the upper atmosphere from a surge in cosmogenic radiation which fed into the lower atmosphere and was taken up by the tree. This had been identified in other records around the globe and dated to 774–5 CE, meaning that the deaths of the trees, and hence the eruption, could be dated to the winter of 822 CE on the basis of the number of rings post-dating the spike. Despite the lack of any analysable tephra shards in the neighbouring Greenland ice core records, there was a very noticeable acidity spike at a similar date to that of the eruption. The identification of this eruption of Katla to such a precise year is all the more impressive as it precedes the settlement of Iceland by fifty years, so there are no corroborative eyewitness accounts.[17]

Mount Hekla may have had some of the earliest precise human-documented eruptions but tephrochronological approaches have also been used in combination with documentary records to improve understanding of several other known eruptions. Occurring soon after settlement, the

MOST UNIMAGINABLY STRANGE

aftermath of the Eldgjá or 'fire gorge' eruption, also from Katla, was record-ed in the twelfth-century *Landnámabók*, but without a clear idea of its date apart from the fact that it was close to settlement times with known settlers close to the source. A range of dates in the 930s CE have been proposed from correlations with acidity spikes from various Greenland ice cores. However, within a 2011 Greenland ice core known as NEEM-2011-S1 it has been possible to detect a clear marker horizon in the 940s CE around which possible acidity spikes could be evaluated. These spikes included one assigned to the Millennium Eruption of Changbaishan, a volcano on the borders of China and the DPR of Korea, that is known to have taken place in 946 CE, again based on the relationship of trees killed by the eruption close to the cosmogenic spike of 774–5 CE and identified by both tephra shards and an acidity peak in the NEEM core. Based on all the acidity evidence and its position relative to the Changbaishan eruption, the Eldgjá horizon suggests an eruption date in the spring of 939 CE.[18]

The impact of the Eldgjá eruption must have been widely felt by the relatively new colonizers of the country and it has been argued recently that the detrimental climatic impacts of such an eruption could have helped hasten a move away from paganism to the adoption of Christianity that took place in the year 1000, especially as foretold in *Völuspá* (The Prophecy of the Seeress) – an important early Icelandic poem. In this poem the end of paganism and the acceptance of a single god is preceded by a series of portents, including descriptions of phenomena very similar to those associated with volcanic eruptions.

> The sun grows dark,
> the earth sinks into the sea,
> the clear, bright stars
> disappear from the sky;
> vapour pours out
> and fire, life's nourisher,
> the high flame
> plays on heaven itself.

At the same time as the Alþingi was debating whether to accept Christianity in the year 1000, there was an eruption on the west of Hellisheiði that produced the *Kristnitökuhraun*, an *aa* lavafield, whose name is variously translated as Christianization lava, conversion lava or Christianity-taking lava. This lava field was the subject of a famous retort by a chieftain called Snorri goði (which means 'high priest') at the Alþingi, as recounted in the thirteenth-century *Kristni saga*. When told of the eruption and its threat to Ölfus, the farm of the pagan Þóroddur goði, the response of the pagan group was to say that it was due to the talk of Christianity angering the gods. In reply Snorri goði, also at that time a pagan, asked what had angered the gods when the lava on which they were standing was burning. This has been seen as the turning point for the meeting to agree to accept the new faith. Whatever the reality of such records, in earth science terms the capacity to use the chemical signature of individual eruptions remains one of the most powerful chronological tools available for recent earth history, and Iceland is uniquely placed to contribute to our understanding of how Earth has changed over this dynamic geological period, especially in the survival of written accounts that may refer to the same events.

On a much longer timescale, the ability to use Icelandic tephras deposited in a range of different sediments to link events across large areas is remarkable. In the marine sediments deposited in the North Atlantic there is a very widespread horizon known as North Atlantic Ash Zone II (NAAZII), a series of volcanic ashes with a clear rhyolitic component. Although there is some dispute over the exact source of this ash it is found on land in Iceland as the Þórsmörk Ignimbrite, originating from under the ice at either Torfajökull or Tindfjöll. Ignimbrite is a pumice-dominated hardened tuff (ash) with the glass shards consolidated by pyroclastic flows, some of the most unpleasant and deadly forms of volcanic activity. K-Ar dating of the Icelandic deposit has given an age of 55.6±2.4 ka, which compares with an ice core age of the same ash from Greenland of 55.4±1.2 ka, an astonishing level of comparison at such a time period, and one that allows NAAZII in the ocean sediment to be assigned the same date, demonstrating the power and potential for eruptions to be used to understand geologically recent earth history and the processes involved.[19]

Although in the previous chapter attention at Þingvellir focused on rifting and the landscape in the immediate surrounds of Almannagjá, the area as a whole has evolved under the influence of four independent volcanic systems. The view to the north from Almannagjá, quite well captured by Josiah Wood Whymper (the father of the Alpinist Edward Whymper) in his sketches for Lord Dufferin, is dominated by Skjaldbreiður, which was formed around 9,000 years ago by a multi-decadal effusive eruption leading to an almost perfect example of a shield volcano. Regrettably not all visitors were impressed by the feature, with Alice Selby in 1931 finding Skjaldbreiður 'a tame affair to have made all the lava that became Þingvellir'. To the south, appearing within a large lake called Þingvallavatn, are examples of much more recent and different volcanic activity, especially the tuff cone of Sandey, consisting of consolidated volcanic ash, that appeared only 1,900 years ago. Just as Skjaldbreiður is a classic example of a volcanic shield and representative of volcanic activity over long periods by relatively slow lava extrusion, so the Sandey cone is the textbook image of a tuff cone. Both probably formed the basis for many of the early sketches of 'a volcano in Iceland' that can be found in accounts and articles published by visitors. Using the usual view of Hekla from the west as representative of a typical Icelandic volcano may not have proved popular, as Henderson so succinctly put it in his description of the mountain, just as always showing Hekla with steam/smoke rising from the summit hardly faithfully represents what most people must have seen.

Hekla is one of only a handful of central volcanoes in Iceland, the core eruption centres of individual volcanic systems, not to be covered by ice. Two of the most active volcanoes, Katla and Bárðarbunga, lie below ice caps, Mýrdalsjökull and Vatnajökull respectively, and even the beautiful stratovolcano at Snæfellsjökull, which is such a feature of Faxaflói bay and can be seen from Reykjavík, is covered by a small, if rapidly diminishing, ice cap. Despite the ice cover, Professor Lindenbrook and his two companions were able to start their journey here in Jules Verne's classic 1864 story *A Journey to the Centre of the Earth*: 'Descend, bold traveller, into the crater of the jökull of Snæfell, which the shadow of Scartaris touches before the kalands of July, and you will attain the centre of the Earth.'[20]

The history of each volcanic system has been gradually pieced together from careful mapping and dating of lavas and ashes, such that over the last 11,700 years it is estimated that about 570 km^3 (137 mi.^{3}) of lava has been extruded onto Iceland and there has been an average of twenty or more individual volcanic events every hundred years, a total of over 2,400 individual eruptions. Although active systems have a quasi-periodicity in that eruptions may be expected with some regularity every so many years depending on the volcano, this is not a precise predictable phenomenon, and studies of methods of volcanic prediction are very much in their infancy. Katla appears to be one of the more regular eruptive centres capable of producing highly explosive eruptions,[21] with a periodicity of between fifty and a hundred years, and like Hekla has produced both basaltic and silicic tephras. Given that the last major eruption took place in 1918 there is always concern over any signs of renewed activity. The 1918 eruption, which started on 12 October, a week before the arrival in Reykjavík of the pandemic Spanish flu,[22] led to a glacier flood moving blocks the size of houses and produced a dense local ash cloud within which people describe not being able to see their hands in front of their faces. A similar intense darkness was reported by local inhabitants experiencing the ash cloud from the 2010 Eyjafjallajökull eruption. There is no reason to believe that an eruption on the scale of 1918, or larger, will not recur; what is uncertain is when.

killed or injured.[26] In 1908, the year after von Knebel's death, his fiancée
Ina von Grumbkov and his friend the geologist Hans Reck revisited the
site but found no signs of the missing explorers. They took the opportunity to travel widely on horseback and develop their own geological
knowledge, as well as naming the crater Lake Knebel, since renamed as the
geographically correct Öskjuvatn (Askja lake). Reck and Ina, who became
his wife, later went on to find the first human specimens in the Olduvai
gorge in Kenya.

Askja had earlier proved a magnet for W. G. Lock, who described it
as the 'most important and interesting' volcano in Iceland. He produced
a book about it that recounted his journey across Ódáðahraun, and then
developed his thoughts on the genesis of Askja and of Iceland in general.[27]
A later visitor, W.S.C. Russell, travelled to Askja in 1913 accompanied by
a Mr J. C. Angus of York, with the intention of surveying the crater. He
reached the summit overlooking the crater lake at midnight with full
midnight sun in the company of a local farmer, Thorður Floventsson, who
was acting as his guide. According to Russell's account he was struck with
Henderson-like awe over the views he saw:

> Midnight! I shall never forget it! . . . The smoking soltafara on the
> eastern shore was sending its columns of steam and gases above
> the crater wall, glistening in the midnight sunshine, reminding us
> forcibly of that other pillar of cloud and the column of fire which in
> ancient days separated the Chosen people from the Egyptian host.[28]

Russell speculated about the demise of von Knebel and Rudloff, commenting that 'the story current in Europe' that the heat of the water in the lake had melted the boat seemed highly unlikely. He favoured an explanation based on his own experience of falling through the snow and ice into a crack within the crater, only being saved by being roped to other members of his party. Had the same happened to the Germans while carrying the boat it could be an answer to the riddle of their disappearance and presumably could account for the lack of any sign of the boat, but drowning is still generally favoured as the likeliest cause. While in the crater Russell took several photographs, producing a panorama of the area, but only selected images are produced in his account. He also completed his survey and could not resist naming features after those who had been involved with the volcano. Thus there are Wattsfell, Rudolfsgygur and Thoroddsenstindur, for example, identified on his sketch map.

The slopes around Askja are also renowned for later visitors: in 1965 and 1967 astronauts from the USA *Apollo* missions to the Moon trained there at Nautagil, a name given from their link to the area. Overall 32 astronauts used the site, seven of whom landed on the Moon's surface, and their experience is now marked by a monument outside the Exploration Museum in Húsavík unveiled by the children of Neil Armstrong in 2015, a museum (yet another) housing memorabilia from the Moon missions.[29] One of the astronauts, Edgar Mitchell, who flew on *Apollo 14*, spent the longest time of any of the missions on the lunar surface and was profoundly affected by his lunar experience. He also commented that the training area 'was a misty, surreal place, unlike anything I'd ever seen on my travels'. Mitchell was so influenced by both the Moon and Iceland that he spent the rest of his life attempting to reconcile his science with a belief in aliens and the importance of awareness-raising, founding an Institute of Noetic Sciences that specializes in the 'intersection of science and human experience'.[30] In the eyes of NASA, he became seen as a bit of an oddball. Manned lunar exploration may have been suspended but the barren surfaces of Iceland continue to provide an experimental outdoor laboratory for NASA. The FELDSPAR project (Field Exploration and Life Detection Sampling for Planetary Analogue Research) uses recently erupted Icelandic lava plains as 'lifeless' surfaces analogous to what might be found on Mars. The aim was to establish just how many sites would need to be sampled in the field by the *Mars Rover* to establish the presence or absence of microbial communities.[31]

Although the lake at Askja may have claimed lives in the last two centuries, Iceland has seen virtually no deaths from volcanic eruptions over the same period, which is remarkable considering that one eruption took place less than half a kilometre from the edge of a town. At 1.55 a.m. on 23 January 1973 a 1.6-kilometre-long (1 mi.) fissure opened up on Heimæy on Vestmannæyjar (the Westman Islands).[32] The source was Eldfell, a developing central volcano in what is a very young volcanic area of eighty separate vents in the Eastern Volcanic Zone; the whole archipelago that comprises Vestmannæyjar was only formed within the last 20,000 years. The magma that was extruded came from the same source as the Surtsey eruption ten years previously, but had evolved slightly due to its storage in a holding magma chamber at a depth of 10–20 km (6–12 mi.). Amazingly the whole island was evacuated successfully within six hours, with three hundred less able residents leaving by plane within two hours and the remaining 5,000 taking advantage of the storm-bound fishing fleet that was in the harbour, supplemented by boats from the mainland. The eruption lasted five months and covered a quarter of the town with lava and ash, increasing the size of the island by 2.2 km² (¾ mi.²), but it did not cut off the harbour as seawater was hosed over the advancing lava to allow it to cool. Within two years around 80 per cent of the population had returned and it remains an important fishing harbour.

There was only one fatality through gas poisoning that could be attributed to the eruption, yet the history of Vestmannæyjar has not always been a happy one. According to *Landnámabók*, it took its name from being settled by Irish slaves, Westmen, who killed Hjörleifur Hróðmarsson, the foster brother of Ingólfur Arnarson, the first settler of Iceland. Around 875 CE Ingólfur Arnarson pursued the slaves and killed them as retribution but the islands were soon resettled, by 900 CE. In 1627 they were subject to *Tyrkjaránið*: raids by pirates in Barbary corsairs from Salé in Morocco and Algiers.[33] They were led by Dutch *renegados*, captains who had converted to Islam and were effectively mercenaries. In all, four hundred Icelanders were captured and a month later found themselves being sold as slaves in North Africa. Two remarkable stories arose from the abductions. Those captured in the Heimæy raid included the Reverend Ólafur Egilsson and his young and pregnant wife Ásta. She gave birth at sea and they were all sold in Africa. Reverend Ólafur Egilsson was released to travel back to Denmark to negotiate a ransom but none was forthcoming and he returned to Iceland, where he wrote an account of his experiences. Eventually his wife, but not her son, was ransomed and she returned, allowing them to be together for three years before her husband died. A fisherman's wife, Guðríður Símonardóttir, was also captured along with her son and sold as a concubine before being released following payment of a ransom by the Danish king, Christian IV. On her way back to Iceland she met Hallgrímur Pétursson, who was studying for the ministry and sixteen years her junior, with whom she had a child. He had been sent to Denmark to work with the Icelanders who had been released, and ceased his training to return with them to Iceland. Discovering that her husband had remarried and had children but had since died, Guðríður Símonardóttir and Hallgrímur Pétursson married. He was eventually ordained and is best known in Iceland as the writer of what are known as the Passion Hymns, *Passíusálmar*, which are sung during Lent.

The Eyjafjallajökull eruption provided one of the first opportunities to observe and measure at close quarters an accessible active volcano in Iceland. The eruption began on 20 March 2010 as a small fissure with little explosive activity in a narrow ice-free corridor between the Eyjafjallajökull and Mýrdalsjökull ice caps, by the Fimmvörðuháls track. This attracted many visitors into Þórsmörk by car, bus and helicopter and the newspapers had pictures of diners at a table just in front of the slowly extruding lava, having arrived by helicopter and been served meals cooked on the lava. On 14 April the eruption entered an explosive phase from a crater below the ice, producing an 8-kilometre-high (5 mi.) ash plume that shut down air travel. As the ash entered the jet stream it moved towards Europe, reflecting the atmospheric circulation patterns of the time, which meant that it not only occasionally split and swung back on itself, but concentrated at different altitudes, hence the uncertainties concerning how safe air travel would be. The London Volcanic Ash Advisory Centre (VAAC), based at the UK Met Office, models ash-dispersal trajectories for Icelandic volcanoes to assess impacts on the airspace around Iceland and northern Europe. The NAME (Numerical Atmospheric-dispersion Modelling Environment) model, first developed to understand the dispersal of radioactive particles from Chernobyl, is used to carry out daily trajectory modelling to prepare for potential eruptions, by injecting hypothetical ash at various Icelandic source coordinates and seeing where and at what flight elevations the ash travels.[34] Following the Eyjafjallajökull eruption, the FUTUREVOLC project was initiated, a cross-Europe collaboration to set up long-term monitoring of active areas in Europe, and there is an Iceland Volcanoes Supersite as part of a global network of Geohazard Supersites and Natural Laboratories.[35] One of the more impressive local effects of the eruption was the infill of Gígjökulslón, the lake immediately in front of the outlet glacier Gígjökull, with an estimated 10 km³ (2½ mi.³) of ash from a sub-glacial flood, transforming an extremely scenic icefall and lake, easily accessible by the dirt road leading to Þórsmörk, into a rather dull sediment-filled basin. In his recent revisitation of images of Icelandic glaciers that formed 'The Glacier Series', based originally on images from 1999, Ólafur Elíasson has an excellent photograph of Gígjökulslón showing the beginnings of a lake refilling within the basin.

Increased monitoring of eruptions since the early 1990s, recording data such as ground deformation and gaseous emissions, has led to an improvement in knowledge about how eruptions have evolved, tracking the intrusion of magma from depth, the propagation of dikes and sills below the surface and eventual surface exposure of the magma. Thus in retrospect it has been possible to provide a better understanding of exactly what happened at Eyjafjallajökull with the emplacement of sills and dikes during the earlier seismic phases, eventually leading to a 4 km (2½ mi.) dike comprising basaltic magma which propagated from the Eyjafjallajökull caldera before emerging at Fimmvörðuháls. The main explosive eruption comprised more evolved magma as it incorporated other material en route to the surface. The experiences gained from monitoring this eruption allowed an even more detailed assessment of the later 2014–15 Holuhraun fissure eruption from the Bárðarbunga volcanic centre under the Vatnajökull ice cap, as well as the identification of the development of a 47-kilometre-long (29¼ mi.) dike leading away from the sub-glacial Bárðarbunga caldera before emerging at the surface at Holuhraun, north of the ice cap.

Research on the 8,000-year-old eruption of Borgarhraun, north of Mývatn, has allowed an estimate of just how quickly magma can travel from the crust-mantle boundary or Moho layer (the Mohorovičić discontinuity) to the surface if not stored on the way.[36] Travelling at between 0.02 and 0.1 ms^{-1} (metres per second),[37] it took ten days to travel 24 km (15 mi.). The magma movement was preceded by gas and it has been suggested that monitoring gas, in this case CO_2, could warn of an impending eruption, although at most the warning would have been two days in advance. Given the improvement in understanding of how volcanoes work it may be queried whether in Iceland it should now be possible to give adequate early warning of impending eruptions. Apart from the danger of any warning leading to apocalyptic headlines in the non-Icelandic press, there is obviously great value in being able to provide some sort of alarm system. Fourteen of the last 21 Icelandic eruptions were the subject of successful warnings; four were detected before being seen and three seen before being detected. Only three had no warning. Given that the period between the first precursors of an eruption, usually an earthquake swarm, and the actual eruption can vary from a matter of minutes to several days and can also include a quiescent phase, it is not surprising that prediction is so difficult. In Iceland, Hekla seems to be the most difficult volcano to predict in this respect.[38] At the time of writing there has been a sequence of earthquakes and the imposition of magma leading to a rise in the surface of 5 centimetres (2 in.) taking place at Þorbjarnarfell very near to the Blue

Lagoon, but the monitoring and regular warnings stress the unlikelihood of a full eruption and visitors continue to flock into the pools for their unique experience of Iceland. The volcanic activity also led to a warning not to explore the lava caves at Eldvörp due to the dangerous build-up of gases.

One of Iceland's most destructive eruptions in terms of loss of life took place over 650 years ago in 1362; its source was the Öræfajökull volcano lying under the southern part of the Vatnajökull ice cap, close to the highest summit in Iceland. The scale of the losses is unknown but an account written two hundred years later stated that 'no living creature survived, except an aged woman and a mare.' The particulars of this account, however, show similarity to other stories in Icelandic fables associated with eruptions, hence what actually happened will remain uncertain, but the eruption and associated floods devastated farmland around the mountain, with farmers not being able to return for forty years. Apart from a later eruption in 1727 that killed three people in floods as the ice melted, the volcano has been considered likely to be extinct, but seismic activity was reactivated in 2017, and in July 2018 it was estimated that 10 km³ (2½ mi.³) of new magma had been injected into the local magma chamber, a volume equivalent to that measured eight years earlier at Eyjafjallajökull. Having experienced the 2010 eruption and recognizing that the 1362 Öræfajökull eruption involved a Plinian-type explosive eruption with an estimated plume height of up to 12,000 m (39,370 ft) and local pyroclastic flows and

Veiðivötn 1477 eruption, rhyolitic Námshraun lava flowing into Frostastaðavatn.

jökulhlaup,[39] the Icelandic Meteorological Office raised the Aviation Colour Code for Öræfajökull, the assessment of potential for damage to aircraft, from Green to Yellow, implying that a disruptive eruption could be imminent but as yet nothing has occurred. The geologist Þorvaldur Þórðarson has also cautioned that, in concentrating on erupted ash and floods, it would be easy to ignore the danger of pyroclastic density currents – mixes of gas and debris travelling at between 100 and 200 kph (62–124 mph) and at temperatures of up to 500°C (932°F) – which in other volcanic regions have led to significant loss of life.[40]

Within Iceland, eruptions have variable effects depending on location. While they can be devastating around Katla, Öræfajökull and on the Reykjanes peninsula, with both ash and lava affecting people directly, other areas are mainly affected by ash falls. The impact of these can diminish rapidly away from their source, and with a predominant westerly air flow many eruptions are not recorded in the north and west. The 1477 eruption of Veiðivötn, a 60-kilometre-long (37 mi.) fissure eruption at the southern end of the Bárðarbunga system, that possibly lasted until 1480 and is considered one of the largest known explosive eruptions in Iceland, saw ash deposited over a 53,000-km² (20,460 mi.²) area to the east, northeast and north-northeast and new features formed along the fissure. Yet because its main effects were felt in uninhabited or thinly habited parts of the country it did not impact greatly on the population. Estimates of the number of Icelandic eruptions over the Holocene vary, despite the earlier figure quoted of 2,400. Between six hundred and seven hundred have some physical record but it has been suggested that there may have been up to 1,900 explosive eruptions in all, with more that had no explosive phase. Sites in south Iceland around Katla and Hekla record the most tephras, with individual sections registering up to two hundred eruptions, and, by comparing the ratio of the number of known historical eruptions with those leaving a tephra record, it has been estimated that over the last 8,500 years alone (excluding the pulse of activity at the beginning of the Holocene) there may have been 1,320 separate explosive basaltic eruptions: 340 from Katla, 560 from Grímsvötn, 350 from Bárðarbunga-Veiðivötn and 70 from Kverkfjöll. As the ratio of basaltic to silicic eruptions is somewhere around 4:1, the total number would be nearer 1,700, supporting estimates that it is more than likely that there have been over 2,000 Holocene eruptions, hence the figure of 2,400 that is usually quoted. In eastern Iceland the maximum number of tephra layers found so far is 157 in Lake Lögurinn, and in the west it is 38 at Haukadalsvatn.[41] In the Vestfirðir region of northwest Iceland, this number reduces to 30.[42] In looking to the future, it is difficult to know whether the next major eruption will be at one of the regularly active sources such as Hekla, or from some relatively new source considered at low risk given our understanding of its eruptive history. Time will tell. What is certain is that there are no reasons to believe that regular volcanic events of varying form and size will not continue.

LAKI AND THE IMPACTS OF ERUPTIONS BEYOND ICELAND

'EVERYTHING I NOW WRITE, ACCORDING TO WHAT I SAW OR EXPERIENCED MYSELF . . . I KNOW BEFORE GOD AND MY OWN GOOD CONSCIENCE TO BE RIGHT AND TRUE.'[1]

OPPOSITE
Aerial view of Lakagígar, source fissure of the 1783–4 eruption.

Travelling from Vík to Kirkjubæjarklaustur along the southern margins of Iceland, one is struck by suddenly moving from the flat, sandy expanse of Mýrdalssandur, the area south of Mýrdalsjökull fed by meltwater from the ice cap, to a strange, monotonous, still flat, hummocky surface covered by grey-green moss. This is Eldhraun, the westernmost lobe of the lava extruded during the Laki eruption from the Lakagígar crater row, the Skaftáreldar Fires, that began on 8 June 1783 and eventually ceased on 7 February 1784. It was the first eruption to be the subject of continuous direct observation thanks to the Icelandic cleric the Reverend Jón Steingrímsson, whose parishioners were severely affected by ash, gas and lava from the eruption. Had his account of the events not been in Icelandic – as *Fullkomið rit um Síðueld* (A Complete Treatise on the Síða Fires), and only printed in 1907; not translated into English until 1998, when it was published as *Fires of the Earth: The Laki Eruption, 1783–1784* – it is likely that he would have been recognized as an important early contributor to our knowledge of volcanic geology. Although he could not see directly the source of the eruption he was able to document fully its sounds and smells, as well as the flowing lava and ash that affected the area and his parishioners around Kirkjubæjarklaustur.

He was not the first to describe a lava flow – that had been done for Mount Etna over a century earlier – but his detailed observations and complete chronicle of events provide a unique introduction to an eruption that had a considerable impact on Iceland and beyond. Over 20 per cent of the population perished in the ensuing *Móðuharðindin*, variously translated as 'Mist Famine' or 'Mist Hazard', as did over 80 per cent of the island's sheep, either directly due to fluorosis and fluorine poisoning or indirectly in the associated famine. The effects of sulphur dioxide in the ash clouds were felt

in Europe, and further afield, when dry fogs occurred leading to illness and in some cases death, especially among those working outside in the fields at the height of the 1783 summer.[2] It has even been suggested that the climate impacts were felt in the summer of 1784 by the indigenous Inuit of northwest Alaska, with famine and population reduction as a result of the coldest summer conditions of the last four hundred years. The sulphate aerosols from Laki remained in the atmosphere for over five months as a veil across the northern hemisphere and caused local cooling of up to 1–3°C (1.8–5°F). Initially in the summer of 1783, as high pressure developed over western Europe, drawing in the ash and gas, temperatures actually rose for a time.

Laki is part of the Grímsvötn volcanic system in the Eastern Volcanic Zone. Although the central volcano lies below the Vatnajökull ice cap, the 1783–4 eruption occurred along an ice-free 27-kilometre-long (16¾ mi.) fissure with 140 separate vents. Grímsvötn is one of the most active Icelandic systems, experiencing at least seventy eruptions over the last 1,130 years, and the Laki eruption was the second largest flood basalt of the last millennium. From the observations of Jón Steimgrímsson and the resultant physical evidence, it has been possible to produce a model of just how the eruption evolved. Magma originated from the crust-mantle boundary and activity occurred in a series of ten major episodes, with earthquake swarms lasting several days and short vigorous explosive events leading to the ejection of sulphur-rich ash into the atmosphere. These episodes were also marked by pulses of increased lava flow following river valleys to the east and west of the mountains that directed lava to the south.

Jón Steingrímsson's parish was centred on Kirkjubæjarklaustur, immediately south of the mountains that hid the vents. He could, however, see the flames and clouds rising into the atmosphere and heard the crashes and explosions associated with each episode. He was also very aware of the lava streams which, flowing within river valleys trending from north to south, brought lava within sight and caused the abandonment of outlying farms. As the lava poured into the valleys the rivers downstream dried up, with the Skaftá taking only two days from the onset of the eruption to become dry, and in a further two days the water was replaced by a 'flood of lava . . . [that] poured forth with frightening speed, crashing, roaring and thundering'. Kirkjubæjarklaustur and Jón Steingrímsson's church lay at the centre of a pincer movement of lava that threatened to overrun the whole area. Matters came to a head on 20 July 1783, the fifth Sunday after Trinity, when with lava from the western arm threatening Kirkjubæ-jarklaustur Jón Steingrímsson held a mass for those parishioners able to attend, believing it to be the last at which he would be able to officiate in the church. He describes how, on leaving the church, they discovered that the lava had effectively halted: 'it had collected and piled up in the same place, layer upon layer . . . and will rest there in plain sight until the end of the world.' This became known as the *Eldmessa* or Fire Mass, Jón Steingrímsson as *Eldklerkur*, the Fire Priest, and the termination of the lava stream as *Eldmessutangi*, the Fire Mass Spit. Although the eruption carried on for several months after this, the church and surrounding houses remained unaffected.

Jón Steingrímsson was born in 1728 and like other priests of his period not only ministered to the spiritual needs of his flock but acted as a doctor over his widely dispersed parish. He would have had a relatively broad training and as an intelligent and curious person was well aware of contemporary writing over a wide range of subjects. It is not known whether he read at all about matters we would now describe within the overall umbrella term of earth science, although he did in later life write an autobiography translated into English in 2002 and published as *A Very Present Help in Trouble: The Autobiography of the Fire-priest*.[3]

His life prior to his time at Kirkjubæjarklaustur had not been easy. Born in Skagafjörður and undergoing training at Hólar, both in northern Iceland, he became deacon at Reynisstaður. When Jón Vigfússon, the priest in charge of the monastery, died, Jón Steingrímsson became friendly with his widow, a woman of considerable talents called Þórunn Hannesdóttir, and they had a daughter 'rather too early'. Jón Steingrímsson was accused of killing her husband and by his own admission in his autobiography had not led a blameless life, dabbling with sorcery while at school and being rather free in his relations with women. He was cleared of the charges against him and moved to Fell í Mýrdal in the south, arriving in time to see at first hand the effects of the 1755 eruption of Katla. Over time his past problems were overlooked and he was ordained, also becoming a widely respected doctor. He was even recognized by the Danish government, who granted him Danish citizenship, which would have allowed him to leave Iceland and live in Denmark, but he stayed in Iceland. Following criticism from some of his parishioners at Fell í Mýrdal in 1779 he moved with his family of three step-children and two daughters to become dean of the whole Skaftafell district. His wife's death at the same time as the eruption in 1784 had such a devastating effect on him that at one stage he considered suicide. He did later remarry but this marriage proved short as he died in 1791. Although perhaps best known for his account of the eruption, his autobiography has been considered the most important comparable work written in the eighteenth century in Iceland, his justification for writing it being to tell his side of his life story and to demonstrate the providence of God in his experiences, especially during the time of the eruption.

View of the western lobe of the Laki lava approaching Kirkjubæjarklaustur showing *Eldmessutangi*, the Fire Mass Spit, in the river channel where the lava stopped on 20 July 1783. There are older rootless cones along the right-hand side of the road and Vatnajökull forms the distant skyline.

Jón Steingrímsson was in no doubt as to his faith and laid the blame for the troubles heaped onto his parishioners as divine retribution in the form of God's Fire on their leading lazy and inconsiderate lives at a time of plenty that preceded the eruption. The halting of the lava flow was to him a crystal-clear demonstration of the power of prayer, as was the way the eruption developed:

> It can tangibly be seen how knowingly and wondrously the omniscient Lord of All Creation directed the course of events . . . from the first minute to the last did merciful God, having decided to destroy a building or farm, grant each and everyone enough time and opportunity to save his life and possessions.

Nevertheless the priest had a very keen scientific eye and did not divorce his need to see the hand of God in all that occurred from making precise and well-founded observations about how the lava behaved. He was the first to describe Pele's Hair: fine, hair-like, glassy threads that form above the flowing lava as the filaments cool and solidify. This was only later named and described in 1849 from observations in Hawaii, using the local goddess of volcanoes for the terminology. He also carried out a simple but effective experiment to determine the relative melting points of the lava and surrounding rocks. He observed that as molten lava flowed down the channels it reincorporated previously cooled lava crusts. In order to show his people that this did not mean that the lava would melt the surrounding mountains, he threw rocks of 'grey stone', clay, earth and sand into the flow and observed how they survived, since they would have needed a much higher temperature to melt: 'I used this result to reassure and convince the people that our mountains . . . were in no danger of destruction by this fire, as people had previously feared.'

Medically he was interested in the effects of the ash and gas on both animals and humans and described graphically the effects of fluorine poisoning as the mouths, nostrils and feet turned yellow and raw, the main reason for the very large loss of life in livestock. The direct effects on plants were also covered, first seen on plants with leaves, then sedges and finally horsetails (*Equisetum* species). As befits a plant group that has survived since the Devonian period, the latter were also the first to recover. Overall his description of the air as having a foul smell 'bitter as seaweed' and 'reeking of rot' provides a very real feeling for what it must have been like for the local community, who were effectively trapped between two arms of encroaching lava with no means of escape or rescue.

As Jón Steingrímsson was documenting the volcanic outpourings threatening his parish, he also realized that the land over which the lava was flowing and the ash was being deposited comprised the deposits from previous eruptions. It was not until the 1940s that Sigurður Þórarinsson formalized tephrochronology. However, it was Jón Steingrímsson who noted that the separate ash layers that could be seen in the soils, particularly as revealed in the extensive eroded faces of *rofabörð* (isolated islands or fragments of the thick loessial soils that covered large parts of Iceland prior to humans arriving – often called rofabards or rofbards in the scientific literature[4]), had to have been deposited prior to the 1780s. He had seen the explosive eruption from Katla in 1755 which deposited ash so it was relatively simple to explain the origins of the ash layers he could see exposed as representative of different earlier eruptions. Although the Laki eruption included several explosive episodes and ash deposits, the extensive lava streams were something he had not experienced. He did, though, realize that in following the river channels the lava was flowing over an earlier very similar lava flow. This was the Eldgjá fissure eruption of the 930s CE discussed earlier, which is considered the largest eruption of historical time in Iceland and originated from the Katla volcanic system, not from Grímsvötn, although the fissure was aligned so that the systems ran into each other. The eruption produced a large outpouring of both lava and H_2SO_4: 19.6 km³ of magma as against 14.7 km³ for Laki (4¾ and 3½ mi.³), but probably less than half as much sulphur. One reported estimate of the size of the Laki lava was that it was equivalent in size to the entire mountain of Mont Blanc, probably calculated by the German geologist Gustav Bischof (1792–1870).

The imprints of the two overlapping flows can be seen at several locations around the margins of the deposits, and a particular feature of the earlier flow in the vicinity of Kirkjubæjarklaustur at Landbrotshólar is an area of rootless cones covering 60 km² (23 mi.²) and partly covered by the 1783–4 lava. These are small craters giving rise to a lunar-like landscape which form as the lava flows into water – either a waterlogged area, a lake or possibly the shallow margin of the sea. Contact with the water leads to steam explosions as the gases are expelled from the lava and the cone is constructed from the expelled tephra; it therefore has no vent and is not an eruption per se. Jón Steimgrímsson described these being formed in active lava: 'When the molten lava ran onto wetlands or streams of water the explosions were as loud as if many cannon were fired at one time.' In 1793, not long after the Laki eruption, Sveinn Pálsson on his visit to the area also observed these features and correctly suggested that they did not originate by primary volcanic activity.[5] Rootless cones or pseudocraters

LAKI AND THE IMPACTS OF ERUPTIONS BEYOND ICELAND

Bárugarðinar at Mývatn. The low ridges to the right are produced by lake ice within the area of pseudocraters. The large feature to the left is composed of larger rocks, *hnullungarður*, broken up by lake ice being driven onshore by northerly winds.

were also observed forming during the early stage of the 2010 Eyjafjallajökull eruption at Fimmvörðuháls but were subsequently destroyed by the
eruption. Around Mývatn in the north they form a characteristic feature
of the local landscape, *Skútustaðagígar*, dating from a second Holocene
phase of eruptive activity around 2,500 years ago when the tuff crater ring
Hverfjall was formed by an eruption under the existing lake. At Mývatn
the interplay between the fine-grained ashes that comprise the craters
and the ice that forms on the lake has produced some distinctive features
around the lake margins. They are known as *Bárugarðinar* or barrier
beaches, series of ridges that are constructed of the loose lapilli scoria, the
ash ejected when the crater forms. When the lake ice is pushed on shore it
bulldozes the ash, with some areas having whole sets of ridges that have
probably developed over centuries if not longer. Pseudocraters have also
been described on Mars and used to confirm the likely presence of water on
the planet at some stage. These early Icelandic observations have rarely if
ever been cited in the academic literature, although the excellent Icelandic
examples are often quoted. Online searches for pseudocraters or rootless
cones may bring up the word 'Viking' but this refers to the *Viking* missions
to Mars rather than any Icelandic source.

Jón Steingrímsson was the second Icelandic priest to observe and
record volcanic activity. He was preceded by Jón Sæmundsson, the cleric at
Reykjahlíð who described the Mývatn Fires, which took place between 17
May 1724 and September 1729.[6] This was a discontinuous fissure eruption,
11 km (7 mi.) long, that originated in the Krafla volcanic system (Mý-
vatnseldar) and produced around 30 km² (11½ mi.²) of lava; Jón Sæmundsson's observations of this in December 1728 constitutes the first known
description of fire fountains: 'fire threw up continuously glowing and
wet rocks that when they hit the ground, they flowed like running water
and burned up everything in their way.' Almost a century later, Ebenezer
Henderson was extremely impressed by what he saw in the Mývatn area.
He recognized the lava from the Mývatn Fires as the youngest in the area,
and used his overall experience of Mývatn and Krafla to muse on whether
records of volcanoes and hot springs can be found in the biblical record,
suggesting that 'few . . . would suppose that any traces of lava are to be
found in the Bible.' In the equivalent of four pages of footnotes, which
themselves have footnotes, he then disabuses any sceptic with a very
wide-ranging set of examples from the Old Testament. Henderson also
goes on to liken the view of Mývatn and its environs to that of the Dead
Sea, perhaps overstating his biblical analogies as a result of his observation
that his work in dispensing copies of the Bible was particularly apt in an
area where the 'Oracles of God are extremely scarce'.

Observing the presence of a persistent blue dry fog that would not be dissipated in Paris in the summer of 1783 while acting as ambassador to France, Benjamin Franklin speculated on a possible volcanic origin and further questioned whether the ensuing cold winter might have been exacerbated by this atmospheric phenomenon. We can now identify the climatic effects of the volcanic aerosols reflecting and refracting incoming solar radiation, implicating Laki in northern hemisphere cooling for at least a year following the eruption, yet to place Laki and Franklin as the origin of the specific mechanisms behind this well-established volcano-climate linkage is not totally correct. Franklin suggested that the dry fogs he experienced probably had a volcanic source, but he was mistaken in assuming that such a low-level phenomenon could have such a significant effect on climate. In presenting his ideas he was beaten by a Frenchman, M. Mourgue de Montredon, who argued along similar lines for a link between volcanoes and climate in a lecture at Montpellier on 7 August 1783. Franklin also not only postulated a volcanic source but speculated on a meteoric or cometary source for the fogs, and his original observations on this made in Paris preceded the onset of the Laki eruption. There had, however, been a submarine eruption that created the island of Nyey (New Island) off the southwestern coast of Iceland in early May, seen by Captain Jörgen Mindelberg on board the Danish fishing boat *Boesand*, and news of this reached Denmark later in the summer, before any information about Laki. In his comments on possible volcanic origins for the dry fog Franklin referred to Hecla [sic] and 'that other volcano which arose out of the sea near that island' as likely sources and would only have heard about Laki in the late summer, as news reached Copenhagen, the colonial capital, at the beginning of September 1783. When he presented his 'Meteorological Imaginations and Conjectures' to the Manchester Literary and Philosophical Society in December 1784 Franklin did not refer to Laki by name and also presented his cometary hypothesis as another possible cause. Nevertheless he did still see the fogs, whatever their cause, as the precursor to the cold winter of 1783–4.[7] What is overlooked is that in 1783 the Icelander Sveinn Pálsson similarly speculated on the link between volcanic activity and cooling of climate, commenting that people believed the earth fires were responsible for the exceptionally frosty winter, but such observations never appeared to have registered outside Iceland.[8] It was also Sveinn Pálsson, in 1794, who first located the Laki crater row that was responsible for the Lakagígar eruption.

We should not underestimate the effects that Laki had on life in Europe over the summer of the eruption and the following year.

Estimates of additional deaths in Britain suggest that 23,000 died from poisoning during the summer of 1783, followed by a further 8,000 during the cold winter that followed. The psychological effects were also significant, especially given the uncertain origin and form of the pollution source. In July an article in the *Leeds Intelligencer* newspaper talked of an 'almost universal perturbation in nature . . . in France people were talking very seriously of the end of the world and churches . . . were unusually crowded'. The ensuing poor harvests and rising grain prices eventually led in 1789 to the French Revolution and some historians have been tempted to see Laki as the initial trigger to the peasant unrest. Intense heat and storms in Yorkshire were considered to presage 'the time of the grand consummation of all things'. The poet William Cowper described the Sun as having 'the face of a red-hot salamander'.[9] One indirect effect of the eruption was to frustrate a visit to the Alps by a later Icelandophile, when the dry fog of the summer of 1783 prevented Sir John Stanley from experiencing the sublime in the form of the Alpine landscape – such an attraction to rich young British aristocrats of a Romantic disposition. Six years later Stanley sailed to Iceland via the Faroes and wrote an account of his travels that became familiar reading for later visitors. The most recent analysis of the climatic impact of the Laki eruption, however, has suggested that the severity of the climatic impact may have been overstated and that an equally severe response could have occurred as a result of natural variability in the climate system, the combination of a warm ENSO (the El Niño Southern Oscillation) and a positive NAO (North Atlantic Oscillation).[10] The dry fogs and the high levels of sulphur that reached Europe were, though, clearly of Laki origin and a recent modelling study has examined the likely levels of mortality that Europe would experience were a similar eruption to take place in Iceland in the future. The figures are quite staggering, with estimated additional cardiopulmonary deaths of between 52,000 and 228,000 due to the increased levels of pollution.[11]

It is easy to overlook just how long it took for news to travel prior to the twentieth century, especially from the perspective of the immediacy of information in the twenty-first. Details of the Laki eruption would have been difficult to ascertain even in Iceland, given the location of the source away from Reykjavík, and news of the eruption took until the beginning of September to reach Copenhagen. Attempts to send a ship from Denmark to Iceland to evaluate the impact of the eruption on Iceland were thwarted by bad weather later in 1783 and it was only in August 1784 that officials were able to report on conditions. They were seriously affected by what they witnessed: 'The tormenting sounds of the poor and agony of hunger, the terrible sight of whittled skeletons and desperate behaviour of people and animals alike shall never pass from my memory.'[12]

Visits by officials and scientists from Denmark undertaken as a result of the eruption provide a valuable picture of the nature of the country in the late eighteenth century. Findings were reported to the Royal Danish Academy of Science and Letters, Kongelige Danske Videnskabernes Selskab, which was founded in 1742 and where there had been earlier interest in the exotic geology of Iceland. As most information was published in Danish, not the more widely read languages of English, German and French which were favoured for scientific communication, these observations on Iceland were not readily disseminated in scientific circles across Europe. Travelling through the area thirty years later in 1814, Ebenezer Henderson wrote of 'consequences the most direful and melancholy, some of which continue to be felt to this day'. His detailed account, stretching to twelve pages, may well have been the first to appear in English, but oddly lacks any direct reference to Jón Steingrímsson. There is a brief comment about a priest rescuing church 'ornaments and documents' before he 'took himself to the western parts of the Syssel [region]', and he notes that the lava stopped less than a couple of kilometres to the west of the church at Kirkjubær, but there is no reference to the priest by name or reputation. From the references in his account, Henderson relies heavily on Bjarni Pálsson's visit in 1794 and on the official report to Copenhagen by Chief-Justice Stephensen in 1785, when reviewing the consequences of the eruption for the colony.[13] The lack of any detailed reference to Jón Steingímsson is also odd given Henderson's ability to see the hand of God in everything around him, and in general to produce a well-informed and thoughtful account of his travels, as we have seen from his reactions at Mývatn. The story of a priest who stopped in its tracks the perils of nature at its worst would surely have been music to the ears of an evangelizing Christian. Indeed at Reykjahlíð he comments on the way that the lava of the 1727 Mývatn Fires stopped flowing within 0.6 m (2 ft) of the wall surrounding the church 'as if inspired with reverence for the consecrated ground'.

Like Sir Joseph Banks, Henderson was an accidental visitor to Iceland. He was born in Dunfermline in Scotland and trained as a boot- and shoemaker; however, he then studied theology and was en route to India as a missionary when he was was delayed in Copenhagen. He settled and became a minister at Elsinore, from where he travelled around Scandinavia before his extended visit to Iceland on behalf of the British and Foreign Bible Society.[14] He was an astonishing linguist, being acquainted with over a dozen languages – from Icelandic and other Scandinavian languages

Henry Adlard, 'Ebenezer Henderson, [Agent of the British and Foreign Bible Society]', portrait engraving from Thulia S. Henderson, *Memoir of the Rev. E. Henderson* (1859).

to Arabic, Mongolian, Turkish, Russian and several other European and non-European languages. He eventually returned to England, where he taught theology and Oriental languages. When travelling in Iceland he was able to converse and read in both Danish and Icelandic, as well as Latin, and had access to accounts written in Swedish, German and French; he must have been the most informed visitor to the country over the whole of the nineteenth century, or indeed since. His section on Laki does, though, seem to rely very little on any discussion with local people or detailed observations of the land he was passing through; it reads more as a restatement of previous accounts.

As is normally the case with Henderson, he uses the event to provide an intellectual challenge to his theology, a 'train of serious thought in every reflecting mind', pitting the 'sceptical speculatist' against the 'more experienced and moderate naturalist' (himself) regarding the incompatibility of such a devastating natural catastrophe with the existence of a 'Supreme Superintending Intelligence'. Henderson sees the deity as using volcanism to regulate the Earth for the 'good of the universal system'. Eruptions released the built-up pressure of 'inflammable substances', necessary components of the globe, and prevented even more catastrophic events which would have had more devastating effects on the Earth and all terrestrial life. Somewhat in advance of the internal debates over the ideas of Lyell and others challenging the traditional ideas of earth science and its relationship with theology, Henderson, aided by the broad sweep of literature that his linguistic expertise provided, must have felt challenged continually to make sense of what he saw, heard or read and reconcile it with his unshakeable faith, finding 'sermons in stones'.[15] How he failed to recognize the unique contribution of his fellow cleric Jón Steingrimsson remains a mystery.

HERÐUBREIÐ, MÓBERG RIDGES AND SUB-GLACIAL ERUPTIONS

'WE SAW A HIGH AND SQUARE MOUNTAIN WHICH OUR GUIDE HAD PRE-
VIOUSLY INFORMED US WAS THE ANCIENT VOLCANO OF HÆRDABREID,
WHICH APPEARED TO US LIKE A LARGE CASTLE.'[1]

At midnight on 19 June 1750, having reached the summit of Hekla after a
fatiguing journey up to their knees in snow, Eggert Ólafsson and Bjarni
Pálsson were rather disappointed to see nothing but ice, with no boiling
springs, fissures, smoke or fire. They turned their attention to the view
and were able to see in the distance to the northeast the isolated table
mountain, or tuya, of Herðubreið. The distinctive shape of the mountain
was to be a landmark for many subsequent travellers. William Lord
Watts, on his first crossing of the Vatnajökull ice cap in 1875, described
it as 'shaped like a pork-pie, crusted over with ice and snow upon its
flattened summit, which rose gradually to a fantastic, ornamented apex
in the centre'.[2] Watts also had the foresight to comment that although its
origins were as yet unknown it would require a thorough examination of
all the strata from the bottom to the top to reveal its genesis. It was only in
the twentieth century that geologists proposed that tuyas were part of a
broader group of features produced during sub-glacial eruptions, which
in Iceland now cover an area of 10–11,000 km^2 (3,900–4,250 mi.^{2}). While
tuyas provided one of the odd landscape formations attractive to visitors,
it was perhaps the more extensive finer-grained palagonites, also produced
by eruptions occurring below the ice, and later oddly sculpted by ice, wind
and water, that have a more widespread expression in the landscape, and
which elicited more response from visitors in terms of their inherent
strangeness and exoticism.

The occurrence of eruptions below ice or water produces a variety
of deposits and forms but until the growing acceptance of glacial theory
and the recognition of earlier more extensive ice cover over the island,
explanations for such features were lacking. While around 20 per cent of
the active volcanic area of Iceland is currently sub-glacial, and more than

OVERLEAF
View of Herðubreið showing
the characteristic shape of
the tuya.

50 per cent of historical eruptions have occurred below an ice cover, during
the height of a glacial episode, an Ice Age, this cover would have effectively
been 100 per cent. Iceland provides a rare opportunity to observe both the
occurrence and effects of sub-glacial volcanism. The Móberg formation in
Iceland is a general term for deposits derived from such activity produced
over the last 780,000 years during the latter stages of the Quaternary (Pleistocene) period, and includes a range of rock types including pillow lavas,
hyaloclastites and cap lavas, which are considered to reflect the eruptive
sequences that occur when ice and magma meet. Pillow lavas are produced
first and as there is no mixing of the magma and melted ice, the outer layer
of the extruded basaltic lava is supercooled and thus protects the evolving
pillow-like shape. This is followed by the hyaloclastites (the prefix *hyalo-*
meaning glass-like from the Greek), formed by explosive magma-water
interactions producing largely unconsolidated or consolidated volcanic
glass in the form of breccias and tuffs (consolidated ash), and the whole
sequence is finally capped by lavas produced sub-aerially above the ice and
water surface once the ice surface is breached.[3]

Herðubreið is an excellent example of a complete tuya formation
sequence comprising all elements of the Móberg formation. It was first
climbed in 1908 by Hans Reck, with the Icelander Sigurður Sumarliðason,
as part of his expedition to Askja to search for Walther von Knebel.[4] At 9 km²
(3½ mi.²) it is not the largest tuya in Iceland (Eiríksjökull is five times larger
with a 750-metre-thick (2,460 ft) lava capping), but as an isolated mountain
close to the heart of the country it is a very noticeable landmark. For the
family at the farm at Grímsstaðir, living an isolated life over 48 km (30 mi.)
from their nearest neighbours and who were visited by Ebenezer Henderson
in 1814, Herðubreið was an essential element in their lives. Without watches
they used the mountain as a day mark, the meridian day mark, one of a
series of features on the horizon surrounding the farm that allowed them
to divide the day according to the position of the Sun. Much more common
in the Icelandic landscape than tuyas are the hyaloclastite ridges known as
tindar and *hryggir*, which lack the final lava capping as the eruptions failed to
break the ice surface. There are over 1,000 such *móbergshryggir* in the country,
varying in length but occasionally reaching tens of kilometres, up to 2 km
(1¼ mi.) wide and hundreds of metres high. The 1996 Gjálp eruption below
Vatnajökull allowed observation of the evolution of a classic *móbergshryggir*
(in the geological literature often termed a '*tindar* ridge'). This occurred over
thirteen days under 600–750 m (1,970–2,460 ft) of ice along a 6-kilometre (3¾
mi.) fissure between Grímsvötn and Bárðarbunga. The eruption took only
31 hours to melt through 600 m (1,970 ft) of ice and produced a *jökulhlaup*
that appeared on Skeiðarársandur (see Chapter Nine), but the final

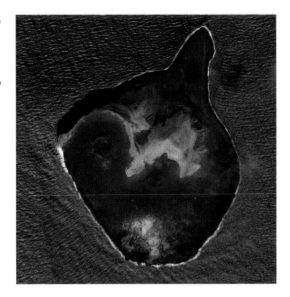

IKONOS image of Surtsey taken on 12 June 2001.

OPPOSITE
Markarfljót, one of the routeways for *jökulhlaup* from Mýrdalsjökull that cut through Late Pleistocene hyaloclastites.

sub-aerial lava stage was very brief. As it cooled it was also visited by helicopter both for scientific reasons and to plant the Icelandic flag on the newly exposed land.

The earlier offshore eruption of Surtsey between 1963 and 1967 provided the first opportunity in Iceland to see phreatomagmatism in action (the interaction between lava and water – although the use of the prefix *phreato-* is misleading as in Greek the term refers to a water well or reservoir, implying a below-ground water source).[5] It also gave the adventurer and author Tim Severin an analogue for the observations made by St Brendan, on his sixth-century ocean voyage, of an island where 'great demons threw down lumps of fiery slag from an island with rivers of gold fire'. Surtsey, like Surtshellir named after the fire-raising giant Surtur of Norse mythology, was one of the longest eruptions of historical time in Iceland, lasting from November 1963 to June 1967. Originating from a vent 400 m long at a depth of 130 m (1,312 and 427 ft), by April 1964 it had built up sufficiently to exclude seawater from the vent. Eventually an island 2.6 km^2 in area and 170 m high (1 mi.^{2} and 558 ft) was produced. Two further islands, Syrtlingur and Jólnir, were formed but disappeared during the course of the eruption due to marine erosion. By 2012 wave action had reduced the island of Surtsey to 1.3 km^2 and the last survey of its height in 2007 was 155 m (½ mi.^{2} and 508½ ft). Submarine eruptions can also produce curious phenomena when they fail to breach the surface of the sea. Eggert Ólafsson and Bjarni Pálsson on their travels in the 1750s referred to an account they had heard that described the sea off the east coast in 1638 turning blood-red and attributed this to the fighting of whales: 'when whales meet and fight . . . and particularly when they are pursued by hundreds of harpoons, the sea becomes tinged with red.' Interestingly, in a different part of the account of their travels, this time about the west coast, they describe similar features that were seen in 1712 and 1649 with fishermen's oars and coastal rocks tinged with red. For one of these they suggest a phosphoric origin from marine organisms, but tellingly for the 1649 event they remark on the fact that it was reported that the previous night the water had appeared to be on fire. We now know that mineral enrichment from eruptions on the sea floor can lead to algal blooms of phytoplankton which turn the sea red or other colours, a feature recently observed, for instance, around Hawaii. The term 'red tide' used to be adopted to describe such phenomena but given the varying colours produced, the more neutral term 'algal bloom' or 'harmful algal bloom' (HAB) is now preferred.[6] Mineral enrichment and blooms of phytoplankton can also be produced from eruptions above the surface that spread ash over surrounding seas. One immediate effect of the spread of ash from the 2010 Eyjafjallajökull eruption was to enrich the North Atlantic with iron, leading to a phytoplankton bloom and what one group of scientists called a 'significant but short-lived perturbation to the biogeochemistry of the Iceland Basin'.[7]

On tuyas with a complete sequence of strata, the height of the boundary between the hyaloclastites and the sub-aerial lava capping provides an opportunity to estimate the thickness of ice at the time of the eruption, as it is considered to mark the highest stand of the englacial lake caused by the melting of the ice above the eruption. In 1965 George Walker used tuyas to map ice thicknesses across north-central Iceland and determine the location of the ice sheet divide. For precise estimates it is necessary to make sure that the tuyas are of similar age, not from different glacial episodes, but using this method it has been estimated that 25,000 years ago at the height of the last glaciation the thickness of ice over Iceland could have been as much as 2,000 m, certainly up to 1,500 m (6,560 and 4,920 ft). Dating of over a dozen tuyas in Iceland using newly developed cosmogenic dating techniques has supported the idea of a burst of eruptive activity as the Icelandic ice sheet disintegrated, with estimates of up to fifty times more activity than later in the Holocene.[8] One of the most distinctive Icelandic tephras to have been identified visually away from the island, the Saksunarvatn ash, was produced by a sub-glacial eruption during the later period of deglaciation. First identified as a visible layer in 1986 in sediments from a site at Saksunarvatn on the Faroe Islands, this tephra has become an important marker horizon across the North Atlantic, with confirmed identifications at 48 locations on Iceland, in 37 North Atlantic marine sediment cores, in Greenland ice cores and across northwest Europe as far as Germany. Rather than a single event, it is now believed to have derived from up to seven different eruptions at Grímsvötn over five hundred years between 10,500 and 9,900 years ago,[9] and is referred to as the G10ka tephra.

Remnants of submarine volcanoes, whose formation follows the same pattern as sub-glacial eruptions, can be seen in the 'islands' of Pétursey, Dyrhólæy and Hjörleifshöfði along the south coast of Iceland. These were submarine volcanoes that emerged and were later joined by the build-up or aggradation of sediment from the extensive river systems of the area, leaving them now above sea level as part of the main landmass of Iceland. Dyrhólæy, one of the best-known rock formations in Iceland, is a classic Surtseyan volcano comprising mainly tuff with a *pahoehoe* lava cap which has been severely eroded since its formation, producing one of the most popular scenic landforms seen by the many tourists who travel along the main road in southern Iceland, which follows the old marine cliffline.

The history of our geological understanding of tuyas and phre-atomagmatic landforms in general is a good example of how geological terminology can proliferate and potentially confuse. In Iceland, after Helgi Pjetursson (1872–1949) at the turn of the last century had recognized the occurrence of a mix of volcanic and multiple glacial events in central Iceland,[10] M. A. Peacock in 1926 described the Palagonite formation of deposits representing this mixing of volcanic and glacial activity. By the 1940s *hryggir* and *stapar* (*stapi* singular) were defined, representing the hyaloclastite or Móberg ridges and table mountains respectively. 'Móberg' is a local Icelandic term, now generally used in the country for what was often called the Palagonite Formation, all strata of a likely phreatomag-matic origin. 'Palagonite' had been coined by the German Wolfgang von Waltershausen based on observations in 1845 at Palagonia in Sicily and was a more widely accepted term in the nineteenth century. Independently in British Columbia in the 1940s the Canadian geologist Bill Matthews was looking at similar features of steep-sided, flat-topped basaltic mountains and adopted the term 'tuya' from a local aboriginal term from the Tuya-Teslin region, for features originating from volcanic activity under a Pleistocene ice sheet.[11] Almost inevitably, given the plethora of terms used over the years, 'palagonite', 'hyaloclastite' and 'Móberg' are often used interchangeably when describing Icelandic features and deposits, and *stapar* is a term still occasionally used despite the acceptance of tuya in wider geological circles. Attempts have been made in the geological

Dyrhólæy, 'Door Hole Island', a remnant hyaloclastite mountain largely formed of volcanic tuff.

literature to standardize the terms used and to give some idea of how to differentiate different forms but effectively the terms '*hryggir*', '*tindar*' and 'Móberg (often given as moberg) ridge' describe elongated ridges made up of a diversity of phreatomagmatic palagonitic material, although the term 'hyaloclastite' is preferred only for rocks with a high glassy or vitric composition. Whichever term is used, geologists should know what is being discussed; it is non-scientists that are likely to be confused.

Had Jón Steingrímsson not had more pressing immediate volcanic concerns, he might have had time to look in detail at the cliffs behind his church in Kirkjubæjarklaustur, for these display a 700-metre-thick (2,297 ft) succession of predominantly hyaloclastite strata with fourteen different lava flows represented. This is known as the Síða Group, which comprises up to 30 km^2 (11½ mi.^{2}) of rock. At the base of the flows are well-jointed lavas, one of which is the Kirkjugólf or 'church floor' pavement described earlier. The friable nature of the majority of the sequences has allowed rivers to cut down extensively through them, especially during the Pleistocene, producing spectacular features such as the Fjaðrárgljúfur gorge, which is up to 100 m deep and 2 km long (328 ft and 1¼ mi.), part of the Katla Geopark. The gorge is believed to have formed very rapidly over the last 9,000 years following the retreat of the last ice sheet, after which a lake formed that was fed by the meltwater driving high volumes of water through the gorge. Eventually the lake infilled, and the river reduced to the size that can be seen today. Having been used as the backdrop for a music

OVERLEAF
Old cliff-line formed largely
of hyaloclastites near
Kirkjubæjarklaustur.

video by Justin Bieber in 2019, it had to be closed to prevent further erosion by followers seeking to visit what is an extremely sensitive site, as are many locations that occur within the palagonite-Móberg deposits.

To the east of Kirkjubæjarklaustur lies Mount Lómagnúpur, the highest cliff in Iceland at 688 m (2,257 ft). Here similar sequences of lavas and hyaloclastites of Pleistocene origin are exposed, forming a further iconic feature along the southern ring road of Iceland, especially noticeable from below Öræfajökull as it appears to mark the western edge of the outlet glacier Skeiðarárjökull and Skeiðarársandur. Not only is the mountain geologically important but it plays a significant role in one of the best-known Icelandic sagas, *Njáls saga* (or The Story of Burnt Njál, *Brennu-Njáls saga*), where it appears in a dream to Flosi Þórðarson at Svínafell, Flosi being the man who orchestrated the burning of Njál's farm. He dreamed that a goatskin-clad giant with an iron rod emerged from the cliff and named the 25 men planning to kill Njál before banging his rod on the ground and returning into the mountain. The giant now forms part of the Icelandic coat of arms. The cliff has also been subject to rockfalls, of which the one following an earthquake and avalanche in July 1789 is the best known. This occurred in the early hours of the day and its remains are well described by Henderson as 'numerous heaps of stone, and immensely huge masses of tuffa, which have been severed from the mountain, and hurled down into the plain'. In all, an estimated 2.5 million m³ (88 million ft³) of rock was detached by the event. The fragility of eroded cliffs comprising principally Móberg rocks in popular areas is now causing concern for tourists' safety. With the recession of Svínafellsjökull below the Öræfajökull ice cap, the Icelandic Civil Protection Agency issued a warning in 2018 for visitors to limit their time and exposure visiting the glacier. A fracture 115 m long and 30 cm wide (377 ft and 12 in.) was found at 850 m (2,790 ft) in 2014 and subsequent monitoring revealed further cracking by 2018, with a 1.7-kilometre-long (1 mi.) fracture system propagating down to the glacier surface at 350 m (1,148 ft). This involves an area of 1 km² (⅖ mi.²) of mountain considered at risk and should failure occur, between 60 and 100 million m³ (2.1 and 3.5 billion ft³) of rock could move, making it one of the largest Holocene earth movements in Iceland.[12] Svínafellsjökull is especially popular because of its accessibility and use as a location for films, including *Batman Begins* (2005), Christopher Nolan's sci-fi adventure *Interstellar* (2014) and more recently the second series of the extremely popular *Game of Thrones*. As there is a weblink as well as an app to guide visitors around various film locations in Iceland it appears almost inevitable that tension will rise between ever-increasing visitor numbers and protection of these locations, which are often geologically sensitive. As we shall see in the next chapter with early visitors to Geysir, this is not a new phenomenon.

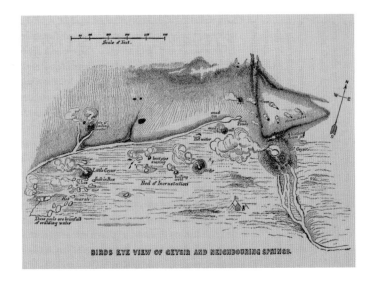

BIRDS EYE VIEW OF GEYSIR AND NEIGHBOURING SPRINGS.

intermittent discharge of water ejected turbulently and accompanied by a
vapour plume' is rather mundane.

The earliest written record of hot springs appearing and disappearing
in Haukadalur following earthquakes dates back to the thirteenth century,
although the name 'Geysir' first appears later at the end of the sixteenth
century, used by Björn Jónsson – the farmer at Skarðsá.[5] The word is derived
from *að gjósa* or 'to gush', generally meaning to erupt, another example
of an Icelandic term becoming part of universal geological nomenclature
having originated with observations by local people. Why the spelling of
Geysir was changed to geyser as yet has no reliable explanation.

Activity at the complex of geysers and hot springs has always proved
erratic. Great Geysir appears to have had quite long dormant periods before
being reactivated by earthquakes. In 1896, following a dormant phase, an
earthquake led to two to three years of eruptions before they tailed off, and
similarly an earthquake in 2000 led to two days of eruptions that reached
heights of 122 m (400 ft) before it effectively became dormant again,
punctuated by a surprise burst in 2016. Currently Strokkur provides a very
regular and predictable display but there is no guarantee that will continue
into the future. In comparing accounts, Barrow found a range of heights for
the eruption from 110 m/360 ft (Ólafsson and Pálsson) to 27 m/90 ft
(Mackenzie). Attempts to trigger displays were made by many, if not all,
visitors, using either a range of objects introduced into the cavities or by
excavating channels into the rim to alter the local hydrology. Soap proved

the most effective stimulant, but rocks and turf were also regularly used. Indeed Cuthbert Peek in 1882 outlines that Strokkur would always be prepared 'to display its powers when properly called on to do so'. This required inserting a barrowful of turf, after which within half an hour it would erupt to a height of 30 m (100 ft). Baring-Gould used turf extensively and also started to dig a channel diverting lukewarm water into Strokkur, but commented critically on the actions of a previous visitor, the German-born professor Carl Vogt. Vogt was a scientist of wide interests who worked with Louis Agassiz at the time of his public pronouncement of glacial theory and after whom a road is named in Geneva in honour of his achievements as professor of natural sciences at the university there.[6] At the end of a tour of the northern seas in 1862, which included landing on the rarely visited Jan Mayen, Vogt arrived in Iceland, where he was feted in Reykjavík. Visiting Geysir he used rocks to encourage an eruption, something Baring-Gould felt had contributed to the inactivity in subsequent years. The criticism of Vogt was a little hypocritical as Henderson had thrown stones into Strokkur, bringing on an eruption within a matter of minutes, and Sir George Mackenzie had done the same, observing that the stones were then ejected to a greater height than the water. From von Troil's record of 1772 in which he observed eleven eruptions in twelve hours, it seems as if he also threw in rocks to see how powerful the jets were. Several engravings used to accompany travel accounts showing Geysir in action have stones being ejected, although they are noticeably absent from the illustration in Baring-Gould's book. In 1834 John Barrow bemoaned the lack of activity at Strokkur, due, he thought, to its being choked by rocks thrown in to encourage a display for the visiting prince of Denmark. He was, though, caught out by a sudden eruption and was lucky to escape without being scalded. It is quite difficult to find any traveller who did not use stones or turf to make sure they saw some action in the brief time available to them at the site. Even as he left Geysir Baring-Gould could not resist one final hypocritical throw: 'Farewell, Geysir! We took one last look into the calm steaming basin, tossed one final load of turf into Strokr, and galloped off.'

Mrs Tweedie recommended that visitors prepare themselves for the occasion with sufficient soap and avoid the local gentleman who tried to charge her 5 Kronur to set the geysers erupting. This sum, equivalent to 5 English shillings around 1900 according to Scott's handbook, was enough to buy a load of turf. The various geysers were also used as rubbish pits for some who threw in chicken bones and Fortnum & Mason jars. Sir Joseph Banks used the hot springs for cooking, as did Lord Dufferin, who favoured locally shot produce and managed to boil a ptarmigan in six minutes; Barrow's curlew and plover took a little longer at twenty minutes.

Mrs Tweedie's fare was a little less exotic: a tin of oxtail soup heated in Blissa (Blesi), the pool that gained the name of the 'traveller's friend' by the 1890s; Blissa's channel was used by William Morris for bathing. As a typical example of the Englishman abroad in the early nineteenth century, George Clayton Atkinson in 1833 put aside any scientific interest to wallow in the comforts the hot springs afforded:

> anyone who holds in due estimation the comforts of a really close shave should go to the Geysir. One uncommonly nice little boiler presents itself to stick his razor in; another beautifully contoured one for his soap.[7]

John Barrow had been requested to find out the fate of a horse reputedly lost in one of the pools by Lord Steuart of Rothsay on his visit in 1831, whose bones were said to have been ejected by an eruption. Although there appeared to be some local knowledge about part of a leg and hoof having been found by the geyser, it was not clear whether the whole horse had disappeared.

The uniqueness of geysers generated genuine scientific interest from visitors trying to understand the processes involved. The link between earthquakes and increased activity, and the recognition of the likely existence of complex patterns of chambers at and below the surface – 'ground honeycombed by disease into numerous sores and orifices', as described by Lord Dufferin – were aspects commented on over many years, but developing a theory to explain the exact process by which water was expelled from the ground proved more intractable. In a letter to his friend the antiquarian Thomas Falconer, Sir Joseph Banks admitted that neither he nor his colleagues could explain why Geysir erupted and that the explanation was that it was 'Lock'd up from the eyes of curious mortals with great care, guarded as she is by fiery dragons'. Credit for establishing a feasible theory is usually attributed to the German chemist Robert Bunsen (1811–1899),[8] whose famous burner has found its way into laboratories across the world. After working on various chemicals, including looking for an antidote to arsenic poisoning, which almost killed him, and losing an eye in an explosion, he and the French mineralogist Alfred Descloiseaux were invited by the Danish government to visit Iceland and review the general geological response to the major eruption of Hekla that had taken place in 1845, following sixty years of inactivity.[9] They spent May–August 1846 in Iceland and climbed Hekla only three months after the complete cessation of the eruption, making various observations. At Geysir they undertook a number of measurements, including taking temperature readings down the

Georg Hartwig, 'The Strokkur [Erupting with Large Stones]', illustration from *The Polar and Tropical Worlds* (1872).

THE STROKKR.

cavity of the spring and letting stones down to different depths, discovering that below a certain depth these stones were not ejected by eruptions. From a later laboratory experiment Bunsen argued that subsurface heating of water expelled air, leading to reduced cohesion and ejection of water and steam. Baring-Gould, considering Bunsen's findings and from his own observations, argued that differential heating within the system would create a downward force on water below and eventually the pressure generated would expel the water and steam nearest to the surface. On his return to England he constructed a bent tube which he heated and managed to produce a water jet that was expelled up to a height of 5.5 m (18 ft), an adaptation of a theory first proposed by Sir George Mackenzie as a result of his visit in 1809. The various theories current in the mid-nineteenth century were neatly summarized by Charles Lyell in his later editions of *Principles of Geology*, where he points out that Bunsen did not propose a uniform theory applicable to both Great and Little Geysir and that much depended on the subsurface structure of chambers and pipes within which water was being heated and hence the temperatures that could be generated. One key finding of Bunsen that Lyell emphasized was that the water involved was originally atmospheric, that is, from rain and snow percolating into the groundwater, not from deep in the crust. Bunsen's observations regarding the colour of the water emerging from the springs, notably their blueness, also helped provide a basis for important discoveries about the scattering of light made

MOST UNIMAGINABLY STRANGE

by the Irish physicist John Tyndall (1820–1893), who studied under him for a time.

Bunsen's ideas are still broadly accepted today, supported by measurements from geysers across the world, as is his recognition that no one theory explains all geysers. As recently as 1972 White and Marler accepted that 'Geysers are exceedingly complex hot springs, no two of which are alike.'[10] Although much early work was based on the Icelandic examples, research in the twentieth century focused particularly on Yellowstone National Park in the USA, supplemented by studies in New Zealand and Chile. Continuing observations of the fewer than 1,000 active geysers around the world is also supplemented by citizen science in the form of the Geyser Observation and Study Association founded in 1983, which publishes, among other things, the *Geyser Gazer Sput*, a bi-monthly newsletter outlining worldwide geyser activity.[11] The use of soap to lubricate geysers had originally been discovered accidentally by prisoners in New Zealand who used hot springs to wash clothes, but this is now considered to be potentially environmentally damaging and at present nature is allowed to take its course, relying in Iceland on the regularity of Strokkur to please the thousands of tourists that visit the site. Measurements at Great Geysir reveal a cylindrical shaft 2–3 m across and 20 m deep with water at the surface at 80°C (6½–9¾ ft, 65½ ft, 176°F) – not that different from observations made by Cuthbert Peek and published in 1882, which he made by letting down a block on the end of a rope. The procedure had to be abandoned rather abruptly as the geyser erupted and they all 'fled away in various directions'. At 10 m (33 ft) depth the temperature rises to 120°C but boiling does not occur below 130°C (248 and 266°F). It is, however, this superheating in the middle section that eventually leads to a detonation and, as the surface rises, to explosive boiling and ejection of water. All geysers may not be the same but heating down the column over cooler water and reduced surface tension appear to be common characteristics. Mrs Alec Tweedie persuaded her father, George Harley, who was a well-known surgeon and Fellow of the Royal Society, to append to her book an explanation of how geysers worked based on his observations at Yellowstone. In this he theorized on a possible mode of action for what he termed in true medical style the 'earth-sod emetic' – why Strokkur erupted when fed with turf.

Descloiseaux speculated that geysers had been present in the Haukadalur area since before colonization and other visitors such as Charles Stuart Forbes in 1860 believed they were a similar age, possibly having read Descloiseaux, who had published his findings in 1849. More recently, using the tephra sequences found in the sequence of

J. Clark, after Sir George Mackenzie, 'Great Jet of Steam, on the Sulphur Mountains', 1811, coloured aquatint published in *Travels in the Island of Iceland, During the Summer of the Year 1810* (1811).

sinter deposits (the porous, vesicular siliceous material that forms an incrustation around the geyser) that comprise the discharge apron around Great Geysir, an age of at least 3,000 years has been proposed for geysers in the area as a whole, concentrated in four distinct phases of activity.[12] Great Geysir was only initiated in the final phase starting eight to nine hundred years ago, after Norse colonization. It is, though, quite possible that geysers in and around Haukadalur have been operating throughout the Holocene since the disappearance of the main Icelandic ice sheet, although no direct evidence for this has yet been published. There is a rather strange conclusion in Tweedie's account concerning the potential impact of hot springs and the thermal areas of Iceland beyond its shores. As she left Geysir she made the point that, despite the terrors and dangers of volcanoes and boiling water and mud, Britain would lose out without them as they aided the Gulf Stream and warmed the air over Western Europe. It is difficult to see where she got this idea from. She was a well-educated and wealthy woman, certainly the only British visitor to Iceland to start her book with 'As the London season, with its thousand and one engagements . . . draws to a close, the question uppermost in any one's mind is "Where shall we go this autumn?"', and then deciding on September in Iceland. Whether this was the geography she had been taught or her own deduction is not clear. Although rich and privileged, and in many ways, apart from her gender, a typical example of a number of British Victorian travellers, she did in fact become destitute on the deaths

MOST UNIMAGINABLY STRANGE

of her husband and father and eventually had to live off her earnings as a
successful traveller, writer and artist.[13]

Geysir may have been the main attraction but the widespread
occurrence of hot springs across Iceland also proved a magnet for visitors.
Von Troil in Letter xxi – 'Of the Hot Spouting Water-springs in Iceland' –
of his *Letters on Iceland*, written in 1774 but based on his visit with Sir Joseph
Banks in 1772, noted that hot springs (*hverr*) occurred away from volcanic
centres, and he provided an account of all the areas of which he was
aware that ran to five pages preceding his description of Geysir. He also
commented on the variations in temperature between the different areas,
which was a precursor for the later identification of high-temperature
fields (temperatures >150°C/302°F at 1 km/⅔ mi. below the surface) and
low-temperature fields (temperatures <150°C 1 km below the surface). Earlier,
in the sixteenth century, the questionable account by Blefken talked of the
restorative properties of an Icelandic spring that 'bubbles forth liquor like
Wax, which notably cureth the French disease which is very common there'.
The French disease was syphilis, and its occurrence in Iceland at this time
has been confirmed from the examination of skeletons from a burial
ground used in the first half of the sixteenth century attached to monastic
remains at Skriðuklaustur in east Iceland.[14] Later visitors seem to have
overlooked these potentially restorative properties.

Although a little disappointed with Geysir as she did not see a major
eruption, Ida Pfeiffer did encounter areas of springs and geothermal areas

with boiling mud and learnt of the fragility of the surface in these areas. Near Krýsuvík, following her pastor guide who went through the surface up to his knees, she also broke through and was fortunate not to be burned. Previously William Hooker had been alarmed by a 'nauseous gust', and in avoiding injury crashed through the surface, necessitating rescue by his companions. Pfeiffer was surprised at how little her guide was concerned by his accident and was further amazed that local people did not make more use of the endless hot water to bathe themselves. Hot spring areas at various locations were visited with good descriptions of solfatara, volcanic steam vents and fumaroles, especially at Laugarvatn on the way to Geysir, at Námaskarð near Mývatn and Krýsuvík on the Reykjanes peninsula. The erratic nature of these features also became apparent to Cuthbert Peek when he visited a lake south of Krafla that Ebenezer Henderson had described in 1814 as a circular pool of black liquid with a central column of liquid being thrown to 6–9 m (20–30 ft); all Cuthbert Peek found was a lake of cold, clear water. Although activity may have been erratic over time, the regular nature of features in real time intrigued Henderson. At Hveravellir, which he recalled as having particularly beautiful incrustations surrounding innumerable apertures, he comments:

> A scene more brilliant and interesting than any ever exhibited on a birth-day festivity. The order they maintained can only be compared to that observed in the firing of the different companies of a regiment drawn up in order of battle.

Henderson was also captivated by the springs, fumaroles and mud pools he saw in what is now the Námafjall high-temperature area at Krafla. His guide originally refused to take him to the area, saying it was unexplored and fearing hidden pools of boiling clay, but Henderson persisted and was rewarded with good displays from the three main jets he called Norð-hver, Oxa-hver and Syðster-hver. The highest jets of what he saw as 'second rank' features behind Geysir were at Oxa-hver, reaching heights of 4.6–6 m (15–20 ft). He was told by locals that the spring owed its name to a cow having fallen into an adjacent spring that was then ejected from Oxa-hver. Although disbelieving of this explanation he does conclude that it would have been 'by no means improbable'. Henderson also comments on the impressive crater lake of Víti produced during the 1724 eruption, one of two crater lakes called Víti in Iceland (the other is at Askja), the name meaning hell, and refers to earlier descriptions of the area by Eggert Ólafsson and Bjarni Pálsson. However, he was never able to assuage the fears of his guide, who appeared terrified by the sights and sounds of the muddy pools at Krafla. In contrast, the boiling mud pools so enthralled Henderson that he stated that 'The boldest strokes of poetic fiction would be utterly inadequate to a literal description of the awful realities of the place.' On leaving Krafla, with full sermonizing gusto he proclaimed:

> Surely, were it possible for those thoughtless and insensible beings, whose minds seem impervious in every finer feeling, to be suddenly transported to this burning region, and placed within view of the tremendous operations of the vomiting pool, the sight could not but arouse them from their lethargic stupour.

The reaction of the guide and possible links with diabolical sources could still be seen in some subsequent descriptions. As late as 1882 at Hlíðanámur John Coles talks of 'columns being projected to the height of about 10 feet, accompanied by such groans, that one could almost imagine they proceeded from some imprisoned demon struggling to get free'.

Despite the visual and existential attraction of hot spring areas, as Sir George Mackenzie commented when visiting Krýsuvík – 'It is quite beyond our power to offer such a description of this extraordinary place, as to convey its wonders and terrors' – the soltafaras, volcanic vents emitting sulphurous gases, also provided commercial opportunities. Sulphur was in wide demand for pharmaceutical purposes as a preservative and disinfectant, for gunpowder and to light fires. Lepers were clothed in sulphur-treated woollen cloth. It is believed to have been mined in Iceland since the thirteenth century and Krýsuvík and Húsavík, lying at the opposite ends of the central active area of the country, saw the development of local sulphur mines. Near Húsavík, the Þeistareykir mine provided sulphur for the gunpowder of the Danish king and Ortelius' map of 1590 includes the annotation 'Fodinae sulphurae præstantissimae' (excellent sulphur mines) to the south of Mývatn, although their precise location is not clear. Concentrations of sulphur varied between 15 and 90 per cent at Krýsuvík, with the Mývatn source exported through Húsavík considered to be the higher quality area; it was either traded crude or with minimal preparation, as in the form of the treated woollen cloth used by the lepers.[15] Visiting the areas mined for sulphur had almost as much of an effect on travellers as seeing geysers. Pliny Miles was clearly shocked by the visual impact of the mines: 'a scene I shall never forget, a literal pool of fire and brimstone'. The success of the mining depended greatly on the operation of the trade monopoly within which Iceland was forced to operate but by the later nineteenth century, with foreign trade having been liberalized in 1855, British traders showed an interest. In 1857 a Mr Bushby acquired rights to mine sulphur in both the north and the south and exploit what was considered a dormant resource. This irked both the French and Americans, as reported by Benjamin Mills Pierce in 1868, who recognized the military importance of having access to sulphur in times of war, especially as the Sicilian supply appeared to be in danger of exhaustion. In 1876, towards the end of his visit, William Lord Watts welcomed meeting up with members of the Sulphur Prospecting Expedition, who were enjoying the sport offered by the river Laxá, having completed their work at Húsavík. Also at Húsavík was Richard Burton, the British explorer more renowned for his exploits in much warmer climes in Africa, who had previously visited Iceland in 1872 on behalf of a businessman interested in the potential that Iceland's sulphur mines had to offer. By 1882 the Icelandic Sulphur and Copper Company, run in a very hands-on way by a Scot, Mr Paterson, was working mines near Kleifarvatn in southwest Iceland and sought to build a tramway to Hafnarfjörður to facilitate exportation. Remains of the mines at Brenninsteinsfjöll near Kleifarvatn can still be seen and a painting of the active Krýsuvík sulphur mine in 1846 by Emanuel Larsen, painted as part of his post-Hekla 1845 eruption visit, is in the collection of the Statens Museum for Kunst in Copenhagen.

Possible extraction and exportation of sulphur as well as kelp had been an attraction for an earlier British visitor to Iceland who spent time at Geysir and eventually produced an account of his travels that Andrew Wawn described as full of 'exuberant geographical description'. This was Sir John Stanley, who had previously suffered in the Alps from the Laki dry fog, and who sailed to Iceland in 1787 and eventually produced an account supplemented by the diaries of his fellow travellers. Drawings and descriptions of Geysir from his observations usually feature in later accounts and suggest that he had no need to 'feed' it to get a suitable erup-tion. Stanley was not initially taken with his visit; he described having 'a voyage full of inconveniences and disagreeable circumstances'. In turn, despite making a good friend of the Danish-Icelandic scholar Grímur Jónsson Thorkelín (1752–1829)[16] when Thorkelín visited St Andrews to collect an honorary degree, and whom Stanley hoped would accompany him to Iceland, he was not much liked in the country and was suspected of passing counterfeit coins. On return to Copenhagen he expressed relief at having left a land lacking in 'civiliz'd societies', yet within a few years his unhappy memories faded and the lure of possible commercial gain, first through sulphur and kelp and then through fishing, generated a renewed appreciation for the country. Despite his reservations he had also collected valuable geological and botanical specimens, depositing the latter at Kew Gardens in the capable and welcoming hands of Sir Joseph

Banks. Not all those from Britain hoping to secure commercial benefits in Iceland were successful. Tweedie recounts the story of an enterprising Glasgow merchant who wanted to show Icelanders the benefits of electric light for combating the long winter darkness, so he chartered a steam yacht, presumably sometime in the 1880s, and went out to provide a demonstration. Unfortunately he chose to do this in August when the long summer days were still apparent and failed to attract any customers.

In modern Iceland geothermal energy now accounts for 25 per cent of total energy production and is the principal source for space heating and especially for heating the large number of swimming pools. The Hengill geothermal field visible from Þingvellir, the second largest in Iceland, has been operating for over fifty years and exploits geothermal energy from a depth of between 1,000 and 2,200 m (3,280–7,218 ft) at temperatures between 320 and 360°C (608–680°F). This is used both as hot water cooled as it is transported to its eventual user and as steam to drive turbines. In an attempt to make more use of geothermal energy, especially given the environmental concerns over expanding hydroelectricity, Iceland's other main power

GEYSIR, HOT SPRINGS AND GEOTHERMAL ENERGY

OPPOSITE
Hot spring at Hengill,
a high-temperature
geothermal area
south of Reykjavík.

source, Orkustofnun, the Icelandic Energy Authority, in cooperation with other Icelandic energy companies, has developed the Iceland Deep Drilling Project.[17] An exploratory well was started through the Reykjanes geothermal field to drill to a depth of 5 km (3 mi.) into supercritical zones at temperatures estimated to be between 450 and 650°c (842–1,202°F). In comparison to the energy output from 2.5 km (1½ mi.) depth this would be an order of magnitude higher but is obviously a major technological challenge. Successful completion of the drilling was achieved on 25 January 2017 when a vertical depth of 4,500 m (14,765 ft) was reached, with a temperature of 535°c (995°F) at the base of the borehole. Utilization of this energy source is still technologically challenging and some years away. One finding was that below 3,000 m (9,840 ft) rocks that were aseismic – that is, stable – became subject to seismic events during the drilling process and studies of flows at this depth, which will determine how hot water can be extracted and used for power, are only just being implemented. There is some concern at developing potentially risky technologies in southern Iceland as in June 2000 the area had two earthquakes of magnitude 6.6 and 6.5, followed in 2008 by one of 6.1 near Selfoss/Hveragerði. Although quite a lot of structural damage was caused, especially in 2000, there were no casualties, largely because it occurred on 17 June, Iceland's National Day, so many people were outside. Exploitation of geothermal areas for energy somehow seems a more positive use of the resource than seeking sulphur for gunpowder for the colonial monarch or as part of nineteenth-century geopoliticking, but presents a major twenty-first-century technological challenge.

CHAPTER 6

GLACIERS, ICE CAPS AND OVERLOOKED ICELANDIC GLACIOLOGY

'THEIR CURVATED APPEARANCE, AND THEIR INCLINATION — AND THE
BARREN PEAKS WHICH THEY EMBOSOM, SUGGEST THE IDEA OF A VAST
FLUID BODY HAVING MOVED FORWARD INTO THE PLAIN, AND CONGEALED
IN THE ATTITUDE THEY NOW PRESENT.'[1]

In its glaciers and ice caps Iceland was seen by some to offer a further
manifestation of hell, although the glittering ice mountains, or *Yökuls*, as
they were often termed, provided a welcome scenic backdrop to the endless
lava plains across which travellers went by foot, horseback or, in later years,
motor transport of variable reliability and comfort. At current estimates,
11,000 km² (4,250 mi.²), around 10 per cent of the country, is covered by ice,
dominated by Vatnajökull, which is over 8,300 km² (3,205 mi.²) in extent.
Previous estimates had put the overall figure at 13,530 km² in 1906, 11,785
km² in 1943 and 11,200 km² in 1980 (5,225, 4,550, 4325 mi.²), and although
there has been a reduction in extent in recent decades the earlier higher
figures probably to some degree also reflect measurement inaccuracies
in the pre-satellite era.[2] Because of the difficulty in accessing most of
the glaciers they were rarely experienced directly in the eighteenth and
nineteenth centuries, although Ebenezer Henderson successfully climbed
Snæfellsjökull in 1815, from where he had good views of Drangajökull and
Glámujökull to the north and believed he could see halfway to Greenland.
As was usual with him he had appropriate words of scripture on hand to
celebrate his achievement, this time from Jeremiah and Psalm 102.[3] It seems
likely that he was the first non-Icelander to view all the major Icelandic ice
caps. Sir John Stanley had made the summit a quarter of a century earlier
and he and his party wrote the names of their mistresses in the snow as 'an
emblem of their purity'. Visitors to the south of the country were aware
of the problems encountered with the extending Breiðamerkurjökull
outlet from Vatnajökull that threatened to reach the sea and cut off the
road, which would have forced them to traverse over the ice – something
that locals already did when the conditions over the surface of the ice were

considered favourable. The main period for which travellers' accounts are available, from the mid- to late eighteenth century to the end of the nineteenth century, coincided with the major extent of ice over Iceland during the Little Ice Age, and experiences of advancing and threatening glaciers would have been commonplace among local inhabitants.

Interest in the glaciers from outside Iceland intensified with the development of mountaineering in the latter part of the nineteenth century, especially by British climbers seeking adventure away from the Alps. The first recorded traverse of Vatnajökull was undertaken by William Lord Watts in 1875, who experienced 'silence, sublime and flawless'. His journey was not without trials as he wore through the moccasins he was wearing, having considered boots were not suitable for the snow and ice. He ended up with an extremely sore frost-bitten left big toe and a sore nose but did not appear to recognize the image painted of him by Sir Richard Burton on their meeting at Húsavík – and reported in newspapers back in Great Britain – as 'reduced to a perfect skeleton, and when first seen was possessed of neither hat, coat, nor boots. Several of his toes were frost-bitten, and his back from the effects of frost, was a pocked mass of raw flesh.'[4] Öræfajökull had been climbed almost a century earlier in 1794 by Sveinn Pálsson. The highest point in Iceland, Hvannadalshnjúkur was climbed by a young adventurous British schoolmaster, Frederick W. W. Howell, in August 1891 after an unsuccessful attempt the previous year, accompanied by a Mr J. Coulthard of Preston and two local guides, Páll Jónsson and Þorlákur Þorláksson, the former of whom had been with Watts on his traverse.[5] Páll Jónsson had a previously unrecorded mountain, a nunatak discovered protruding above the ice on the crossing, named after him by Watts as Pálsfjall – one of the few Icelandic topographic features with a personal name; after this he was known as Páll Jökull Pálsson. At Hvannadalshnjúkur Howell was alone in climbing the final steep ice wall, having the only ice axe, although his companions considered the ascent too dangerous and appeared happy to allow him the honour. Howell was later to cross Langjökull in 1899 and take some of the earliest photographs that still exist of various locations around Iceland, including several of glaciers,[6] before drowning on a river crossing in northern Iceland. He is buried at Mikibær, Varmahlíð, where his gravestone simply states that he was 'called to his rest from the Héraðsvötn River on 3rd July 1901, aged 44'.

A small number of glaciers are mentioned by name in sagas and other early documents, including Eyjafjallajökull in *Njáls saga* and Snæfellsjökull in *Landnámabók* and *Eiríks saga rauða* – and as Snjófell in *Ármanns saga*. Maps failed to show any glaciers until 1539, when Olaus Magnus used the term *Iokel* over the area of Snæfellsjökull shown as *Snauel*

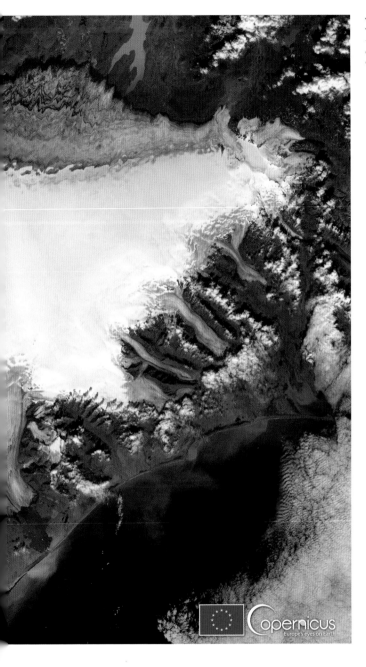

Vatnajökull ice cap, 26 August 2020. The contrast between the darker ice of the ablation area and the white snow cover of the accumulation area can be clearly seen.

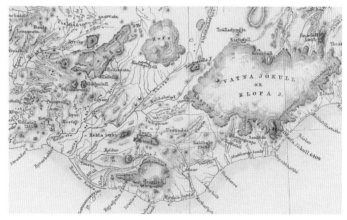

iokel by Ortelius in 1570 in his map of Nordic countries, *Septemtrionalium regionum descrip*, that preceded his Iceland map of 1590. The 1590 map showed eight glaciers – Bald Iokul, Getlands Iokul, Arnafelds Iokul, Sand Iokul, Sneuels Iokul, Mydals Iokul, Eyjafialla Iokul and Solheima Iokul – but did not include any further east than Mýrdalsjökull, although there is a river called Iokuls a in the location of southern Vatnajökull. The important work by Sveinn Pálsson in 1795, whose observations and theories about glaciology will be discussed in detail later, included four maps of the major Icelandic ice caps, Vatnajökull (termed Klofajökull in early accounts), Langjökull, Hofsjökull (Arnarfellsjökull, a name used widely by visitors) and Eyjafjallajökull, which also covered Mýrdalsjökull. He also included many specific names of outlet glaciers from around the ice caps. Building on the comprehensive map of Iceland produced by Björn Gunnlaugsson in 1844, which identified the position of the main ice caps, the first attempt to produce an inventory of Iceland's glaciers was undertaken by the Icelander Þorvaldur Thoroddsen in 1892, published as *Jøkler i Fortid og Nutid* (Glaciers Past and Present). In a series of subsequent publications over the next two decades he refined and expanded his inventory, aided by the countrywide mapping undertaken by the Danish Geodetic Institute that was begun in 1902 and continued until the German occupation of Denmark in 1940.[7]

Glaciers and ice caps in Iceland are generally grouped into eight regions, closely related to the broad drainage structure of the island: Vatnajökull, Mýrdalsjökull, Hofsjökull, Langjökull, Snæfellsjökull, Vestfjarðajöklar, Norðurlandsjöklar and Austfjarðajöklar. These vary in form from the extensive ice cap of Vatnajökull to the over one hundred small glaciers that comprise Norðurlandsjöklar, centred on the Tröllaskagi peninsula. In 1795 Sveinn Pálsson was able to identify 32 place names of glaciers in the latter area despite their scattered distribution and small sizes. Monitoring and mapping are now achieved through using various forms of satellite imagery, with the first analyses published in 1974; echo sounding to produce estimates of ice thickness began two years later.[8] By the year 2000, 269 named glaciers were considered to be active, of which fourteen were classified as individual ice caps, 55 as cirque or corrie glaciers, 73 as mountain glaciers and five as valley glaciers. The southern margins of Vatnajökulsþjóðgarður (Vatnajökull National Park) were defined on the basis of the 1998 ice margin in a move to defuse possible opposition from local landowners. Considering that some felt that they also owned parts of the glacier, and that with subsequent ice retreat defining this boundary on the ground has become difficult, this approach could be considered as less than successful but poses the question of how to identify meaningful boundaries on the ground when physical lines are by no means constant.

Credit for the development of our scientific understanding of glaciers is usually attributed to scientists who based their observations in the Alps, in particular Louis Agassiz, James Forbes and John Tyndall. This is generally given the umbrella term 'glacial theory' and was the subject of a series of acrimonious debates in the mid-nineteenth century between Agassiz and Forbes and particularly between Forbes and Tyndall, who championed alternative theories of how glaciers moved.[9] Yet had it not been for the relative isolation of Iceland, its position as a Danish colony

Winter view of Múlajökull, an outlet glacier from Hofsjökull.

OPPOSITE
Gljúfurárjökull, one of the largest glaciers found on Tröllaskagi.

and the production of written accounts in languages not accessible to the 'mainstream' science of the day, Icelanders whom we would now consider scientists, yet who also had other occupations to follow, would have been seen as important figures in the development of glacial science. The first of these was Þórður Þorkelsson Vídalín (1662–1742), a scientist very much ahead of his time who sought to observe and base his ideas simply on what he saw and could deduce in a logical manner.[10] After a short time as the headmaster of a school at Skálholt, the principal bishopric of Iceland, he became the chief physician for a large area of southern Iceland, which meant he would be very well acquainted with the southern margins of Vatnajökull; hence his work directly referenced Skeiðarárjökull and Breiðamerkurjökull. He talked of 'sheer cliffs of ice that tower over everything, between them innumerable deep chasms', a good description of the highly crevassed marginal areas of the glaciers he visited. His thesis, entitled *Dissertationcula de montibus Islandiae chrystallinis* (Short Treatise on the Ice Mountains of Iceland), was produced in 1695 and preceded writing on glaciers elsewhere. It was not published until it was translated from the Latin into German in 1754 by Páll Bjarnasson Vídalín, but it appears to have been little read. His theory of glacier movement caused by water freezing in crevasses, expanding and leading to winter advances, the frost expansion theory, although incorrect is usually credited to the Swiss scholar Johann Jacob Scheuchzer (1672–1733) in 1705, and the idea was still supported until the mid-eighteenth century.

Þórður Þorkelsson Vídalín did not believe that snow would transform into ice through firn, the crystalline or granular snow formed as it is consolidated and made more dense by the addition of further

snow, and propounded an established view that glaciers were mixes of water and saltpetre recharged from subterranean vaults. In this he went against the understanding of local farmers whose families had lived in the shadow of the ice cap for generations and as Sigurður Þórarinsson pointed out when summarizing knowledge of Vatnajökull in the 1960s, the locals knew more about the glaciers of Vatnajökull, having lived with them for eight hundred years, than could be learned from foreign books. Þórður Þorkelsson Vídalín did, though, recount the local view that Vatnajökull had probably been crossed previously by Jón Ketilsson, who had ascended the ice at Skaftafell and travelled far enough to see the northern margins of the ice cap. Grímsvötn, as represented by a depression in the ice surface, was known of before 1600 but could not have been seen from positions off the glacier. Farmers travelled in both directions across or round the ice cap, from the north to the south for fishing and collection of birchwood, and from the south to the north for collecting hay and Iceland moss. Advancing ice was known to have cut off the easterly route to Möðrudalur by 1575. There was therefore a wealth of local knowledge about the ice cap and its history that was utilized not only by Þórður Þorkelsson Vídalín but subsequently by later scientists. Despite inaccuracies, Þórður Þorkelsson Vídalín's methods and observations provided a template for a more important successor who similarly spent much of his working life as a physician covering a huge area of southern Iceland, bringing him into contact with the glaciers of the area: Sveinn Pálsson.

A man with the looks of an unsuccessful pugilist rather than a scientist and doctor, Sveinn Pálsson (1762–1840)[11] was born the son of a fisherman in northern Iceland, and despite not having the comforts of a private income and/or university position like Agassiz, Forbes and Tyndall, was half a century ahead of these scientists who have become synonymous with the development of glacial theory, the how and why of glacial movement; or as Hans Wilhelmsson Ahlmann later succinctly put it, 'the problems which decide the lives of glaciers'.[12] After working as a fisherman himself following his graduation in 1782 from Hólar, the northern bishopric and educational equivalent of Skálholt, he moved to Reykjavík to study medicine. This took him to the colonial capital of Copenhagen, where he also studied courses in natural history. In 1789 he became a member of the Naturhistorieselskabet, the Society of Natural History, which granted him money to return to Iceland and travel between 1791 and 1794 to produce an account of the natural history of the island in the aftermath of the devastating Laki eruption. Having no further money on the completion of his travels he married Þórunn Bjarnadóttir, the daughter of Bjarni Pálsson, the chief medical officer who had travelled with

Unknown artist, portrait engraving of Sveinn Pálsson, 7 July 1798.

Eggert Ólafsson in the 1750s. He returned to being a farmer and fisherman, but in 1799 was appointed as physician for a huge area of southern Iceland centred at Syðri Vík, stretching from the Reykjanes peninsula to Skeiðarár- sandur and including Vestmannæyjar. He fathered fifteen children, seven of whom died in infancy, and to make ends meet continued his fishing and farming, working until he was 72. Despite all the distractions of his work and family Sveinn Pálsson managed to continue his scientific observations, fighting a losing battle to get his written work to a wider audience.

Arising from his travels supported by Naturhistorieselskabet, Sveinn Pálsson produced a treatise in Danish entitled *Forsög til en Physik, Geografisk og Historisk Beskrivelse over de islanske Is-biærge. I anledning af en Reise til de fornemste deraf i Aarene 1792–1794* (An Attempt at a Physical, Geo- graphical and Historical Description of the Icelandic Glaciers. Based on Travels to the Most Important of Them during the Years 1792–1794). This was submitted to the Naturhistorieselskabet in 1795. The work included maps of the four principal ice caps, Vatnajökull, Eyjafjallajökull (including Mýrdalsjökull), Hofsjökull (Arnarfellsjökull) and Langjökull, and put forward a series of observations and theories that not only questioned the earlier work of Þórður Þorkelsson Vídalín but preceded aspects of glacial theory as exemplified in the debates of Forbes, Agassiz, Tyndall and others by half a century. In contrast to Þórður Þorkelsson Vídalín he accepted that snow transformed into ice at high altitudes and distinguished between

GLACIERS, ICE CAPS AND OVERLOOKED ICELANDIC GLACIOLOGY

perennial snowfields and glaciers, the latter having transformed into ice, before outlining a classification of different glacier types that was very like modern classifications and differed from the previous classification of Eggert Ólafsson and Bjarni Pálsson, despite their objective having been to 'observe the glacier with the most scrupulous accuracy'. Sveinn Pálsson divided glaciers into two groups, summit glaciers (firn glaciers) and descending glaciers (ice glaciers), with further subdivisions based on size and shape, producing a classification that would not appear out of place in relatively recent textbooks. His fundamental separation, though, was effectively an identification of the accumulation zone of glaciers where snowfall exceeds any melt – his summit glaciers – from the ablation zone where losses exceed accumulation – his descending glaciers. While accepting that it was possible that summit glaciers could reach the lowlands due to being forced down during a volcanic eruption – it must have been difficult not to implicate volcanic effects given his experience in the 1780s – he did not acknowledge any uniform theory of how glaciers formed, whereas today we accept how snow is transformed into ice moving from the higher accumulation zone to the lower ablation or melting zone. In this he was echoing the views of the locals that Þórður Þorkelsson Vídalín had ignored, and although there is no direct evidence of any discussions with them about the topic in his diaries, he did stay with Jón and Eiríkur Einarsson at Skaftafell in July 1793, where he commented on their advanced level of learning, reading Greek, Hebrew and Latin, understanding German and Danish, and having interests in surgery and botany. Such were the remarkable capabilities of some of the farmers living in Iceland in a challenging landscape through a time of poor climate and serious volcanic impacts. Sveinn Pálsson was acutely aware that glaciers moved, and suggesting a mechanism for how that took place was one of his most important observations.

In examining the margin of Breiðamerkurjökull, Sveinn Pálsson likened the behaviour of ice to that of pitch (tar) which, although appearing solid, can move and if placed on an incline will flow slowly, yet may also appear brittle. From the summit of Öræfajökull he noted the apparent 'fluidity' of the masses of ice he could see around him. In support of this he cited evidence that crevasses were known to open up and then disappear: 'we have found a new and contributory cause of the formation of many icefalls and creeping glaciers, as well as the disappearance of glacier crevasses in a short time.' A similar idea had been produced independently by the Frenchman André-César Bordier in 1772 but had attracted little attention and his work was almost certainly unknown to Sveinn Pálsson. The recognition of plastic deformation as a major process in how glaciers move is now accepted knowledge but it took almost a century for glaciologists to develop ideas similar to those of Sveinn Pálsson and Bordier. In 1859, when Professor James Forbes was made the first Honorary Member of the Alpine Club, he was lauded with being the 'chief author of the revolutionary view of the nature and movements of glaciers embodied in his famous Viscous Theory', yet the basic idea behind this had already been proposed by Sveinn Pálsson over sixty years earlier.

A related glacial feature to which Forbes has been given the credit for identifying and explaining was ogives. These are alternating dark and light bands that curve down glaciers and predominantly occur below ice falls, representing annual movement through the ice fall. They are sometimes termed Forbes Bands and textbooks describe them as first having been identified by Forbes in the mid-nineteenth century, although the term 'ogive' was coined earlier by Louis Agassiz. Yet Sveinn Pálsson had seen and described these features on Fjallsjökull while climbing Öræfajökull:

> Its entire surface seemed to be covered with curved stripes, extending right across the glacier . . . the curves pointing towards the lowland just as if this icefall had slid down in a halfmelted state or as a thick semi-liquid material.

Although normally, but not exclusively, found below icefalls, excellent ogives on Morsárjökull to the west of Skaftafell provided a focus for later research in the middle of the twentieth century. The lighter bands tend to have relatively clear bubble-rich ice while the darker bands have more massive ice, and close inspection reveals bands-within-bands. The general explanation for their formation relates to compression and deformation of the ice going through the ice fall, but even in 2018 a paper in the *Journal of Glaciology* began with the sentence: 'Ogives remain one of the most enigmatic features of glacier flow.'[13] The paper makes no reference at all to the observations of Sveinn Pálsson, despite his work now being available in English. Ogives should really be known as Pálsson Bands.

Sveinn Pálsson's ascent of Öræfajökull was remarkably rapid, reaching the summit of Sveinsgnípa, at 1,927 m (6,322 ft), in six hours. He was possibly driven to undertake such a climb not just out of intellectual curiosity but by hearing of Alpine first ascents, for example of Mont Blanc in 1786, while studying in Copenhagen, seeking to show he could emulate the exploits of his fellows from mainland Europe. He was accompanied by two others and equipped with a barometer, thermometer, pocket compass, pickaxe, glacier cane and a length of rope measuring eight fathoms.[14] Apart

from noting the ogives, Sveinn Pálsson also heard the explosive sound of
jöklabrestur, or glacier thunder, as chunks of ice collapsed above Kvíár-
jökull. The name 'Sveinsgnípa' was given to the peak in his memory in the
twentieth century. He also climbed Eyjafjallajökull in August 1793.

Of equivalent importance for glaciology was Sveinn Pálsson's
recognition of the significance of convection in the ablation or melting of
temperate glaciers such as those found in Iceland. He argued that solar
radiation would be reflected back by the white glacier surface and be less
effective in melting ice than overcast misty conditions. This theory was
tested on Vatnajökull in the 1930s by Ahlmann, whose results supported
Sveinn Pálsson's ideas, and subsequent studies from other glaciated regions
have similarly provided support, with up to half the ice loss being from
convection. Having walked over ice, Sveinn Pálsson described dirt cones,
pyramids of sand and gravel rising above the surface, and showed them to
be ice-cored. He also provided the correct explanation for 'elf holes', described earlier by Eggert Ólafsson, where dark stones become embedded in
the ice surface, warm up during the hours of sunshine and later radiate out
the heat. This melts the surrounding ice, creating an expanding hole into
which meltwater flows. The stones and material found on the ice surface –
known as supraglacial debris and forming the impressive medial moraines
that can be seen on Breiðamerkurjökull – was recognized by Sveinn Pálsson
as having fallen on to the ice from surrounding mountains higher up the
glacier. Importantly he also suggested that some of the debris was brought

up through the ice from the glacier bed, reflecting his overall view of the plastic nature of ice and its ability to move and transport material.

So why has Sveinn Pálsson not had the recognition he clearly so richly deserved? When he sent his manuscript to Copenhagen it was not prepared for publication, possibly because the members of the Naturhistorieselskabet failed to appreciate its importance, although the decision not to renew his funding after completion may indicate other reasons why society ignored the work. The society itself ceased to exist after 1804 and the treatise was sent on to Norway, obviously in its original Danish form; here it was locked away and forgotten. Sveinn Pálsson tried to get the Literary Society of Iceland (Hið Íslenzka Bókmenntafélag) to publish it in 1832 but was unsuccessful, although his accounts of his travels and other manuscripts were bought from his family after his death by the Icelandic poet Jónas Hallgrímsson.[15] Odd excerpts from the treatise were published through the efforts of Þorvaldur Thoroddsen, who recognized the importance of his work when carrying out his own research in Iceland at the end of the nineteenth century, and Henderson had been aware of his work on his travels, but it was not until 1945 that an Icelandic translation finally appeared – the work of Jón Eyþórsson and Pálmi Hannesson, the former an important figure in twentieth-century Icelandic glaciological research.

Sveinn Pálsson never therefore gained any international credit for the observations and insights he made under extremely trying conditions, although he was recognized within Iceland. The poet Bjarni Thorarensen,[16] a friend of Jónas Hallgrímsson, described him as 'A mind so free and fertile . . . as you [Pálsson] sought to commune with the souls of sages'. Given the availability of his work in English it is sad that current glaciological textbooks still fail to reference his findings, even if Forbes, Tyndall and Agassiz may have reached their own similar conclusions independently. The later publication trials of Watts, who found that proofs of his account of his Vatnajökull crossing that he had immediately sent to *The Times* (with the assistance of Burton) ended up in the wastepaper basket, seem minor in comparison. Watts's observation that 'There are anomalies in the civilised world which confound one even more than the idiosyncrasies of nature' would have struck a chord with Sveinn Pálsson. Louis Agassiz's comment that 'Every great scientific truth goes through three stages. First, people say it conflicts with the Bible. Next, they say that it had been discovered before. Lastly, they say they always believed it' may not have been so warmly received by Sveinn Pálsson had he lived long enough to hear it.

GLACIERS, ICE CAPS AND OVERLOOKED ICELANDIC GLACIOLOGY

CHAPTER 7

MEASURING GLACIERS

'ARE YOU GOING TO VATNA JÖKUL? . . . WELL YOU WON'T COME BACK
ALIVE, NEITHER YOU NOR YOUR DOGS: VATNA JÖKUL IS THE MOST
HELLISH PLACE ON EARTH.'[1]

Interest in Iceland's glaciers during the later nineteenth century from outside the country, especially among anglophones, tended to centre on ideas
of exploration, climbing and seeing areas as yet unvisited. The Icelander
Þorvaldur Thoroddsen (1855–1921) was similarly influenced; noting that
'3,500 square miles of the ice-mountains had never been trodden by human
foot,' he set himself the task of producing a comprehensive geographical
and geological survey of the country, including a more accurate survey of
the glaciers of the country than any previously completed. This involved
seventeen years of fieldwork all over the country. For the first nine years,
while working as a teacher, he was severely limited in his resources but
following the award of a pension by the Alþingi in 1899 he was able to work
from Copenhagen, having studied also in Leipzig, and produce a series of
magisterial publications including a geological map of the country in 1901.
He was very aware of the lack of a scientific infrastructure in Iceland at the
time, a country 'destitute of scientific institutions and laboratories', and
his work was principally published in Danish and German, although he
wrote an overview for the Royal Geographical Society in London in 1899
which was translated into English.[2]

Thoroddsen's work provided the basis for joint Swedish-Icelandic
studies that began direct measurements on the ice to seek to understand
the relationship between glaciers and climate and how they grew and
decayed. The main driver behind this work was the Swedish glaciologist
Hans Wilhelmsson Ahlmann (1889–1974),[3] a professor of geography who
had experience of glaciological fieldwork in Svalbard and Norway
in 1924 and 1925 and who, while working in the Jotunheim mountains
in Norway with the Icelander Jón Eyþórsson, hatched a plan to visit
Vatnajökull. This took a decade to come to fruition but in April 1936 they

Photograph of Hans Wilhelmsson Ahlmann (left) and Jón Eyþórsson (right) in 1936 on the Swedish-Icelandic Expedition taken by L. Ahlmann.

set out with a small group that included a young Sigurður Þórarinsson and four others to climb onto the higher eastern and central areas of the ice cap. There they undertook measurements that would allow estimates of how much mass is accumulated and lost annually on the glacier, the idea of a glacier's mass balance that determines whether it will grow or shrink. Given the challenge of acquiring such measurements, which necessitated digging pits into the snow and firn cover and driving in stakes to observe surface losses, such measurements were relatively rare and hence any data would prove of real scientific value.

Travel onto the ice cap was aided initially by the use of horses and also dogs used to pull sledges, including *pulkas*, a form of sledge used in Lapland. In using horses they followed in the footsteps of Alfred Wegener, primarily remembered as the architect of continental drift, who in 1912 used a stopover in Iceland to test out horses for a crossing of Greenland and completed an eight-day trip over Vatnajökull.[4] Wegener subsequently took 25 Icelandic horses with him for the successful crossing. He died on a later visit to Greenland in November 1930 while taking supplies to a party of researchers on the ice; his body is slowly being transported westwards both on the ice and, perhaps more appropriately given his reputation, on the North American plate. While the human members of Ahlmann's expedition proved trouble-free, one of the dogs, Bonzo, managed to escape and diverted their work as they felt the need to catch him before he headed back off the ice and chased any sheep; financially

Frederick W. W. Howell, photograph of horses crossing the margin of Breiðamerkurjökull, *c.* 1900.

compensating any farmer for the loss of sheep would have been ruinous for the expedition.

Bamboo stakes were driven into the ice in the lower ablation zone below the snowline and regularly observed to see how much melting had taken place. Once above the snowline in the accumulation area of the glacier the group began to dig pits to establish how much snow had accumulated and in particular to look for variations across the different parts of the ice cap. Soon after they started on 26 April they experienced a serious storm with high levels of snowfall. After this abated they continued their digging and had a new marker layer to dig down to. This they termed the tent surface, that is, the surface onto which they had pitched their tents at the start of the storm. Ahlmann was astonished to find that between 26 April and 15 May there had been 550 mm (21½ in.) of precipitation, slightly more than the annual precipitation for his home of Stockholm; they also discovered from a pit below the snowline on Heinabergsjökull that almost 4.5 m (14¾ ft) of snow must be lost or ablated in a year. One of their principal aims was to find the depth of snow that had accumulated above the snow line since the eruption of Grímsvötn in March/April 1934 which had carpeted the ice with ash. They discovered the layer that marked the end of the previous summer at almost 5 m (16½ ft) depth and eventually found the Grímsvötn ash at 8.9 m (29¼ ft). This last feat was much aided by the efforts of one of the local members of the party, Jón from Laug, who was recommended to Ahlmann as being 'as strong as a bear', a devotee of *glíma* wrestling, the special Icelandic version of the sport, and whose prize possession was 'his enormous sleeping-bag made of ice-bear-skin from Greenland'. Finding the ash layer was celebrated at over 1,200 m (3,940 ft) by a tot of Swedish brännvin and a cigar as well as their normal meal of pemmican and Marmite.

As they began their return, Jón Eyþórsson spent a day franking letters with a special Vatnajökull rubber stamp he had been given by the director of the Icelandic Post Office as the designated postmaster for the ice cap, a way to raise funds to support the expedition. In digging the pits they were also able to observe the layered nature of the snow, especially what Ahlmann termed blue snow, where rainy weather had led to water freezing as a layer within the less compact snow. Throughout the expedition Jón Eyþórsson took regular meteorological observations and temperature and density readings within the pits. The temperatures produced surprising results as they varied with depth, with higher readings sometimes occurring below colder ones, especially around the blue snow layers, which remained at 0°C (32°F). On examining the pits Ahlmann felt he was looking at what he considered to be a cryptogram to which he had at last found the key, and he wrote at length of the satisfaction such discoveries offered: 'It does not harm Nature, for Nature only becomes richer and more full of meaning for him the more secrets he finds in her and the more deeply he explores her territories.'

While the gendering of both the scientist and nature may seem dated, the former does accurately reflect the nature of his profession for the pre-War period, and after. Later field measurements of mass balance undertaken between 1993 and 1998 revealed even higher amounts of snow accumulation of between 10.2 and 13.2 m (33½–43¼ ft) in a year on the highest plateau, two to three times as much as that found on the northern and western parts of Vatnajökull.

Ahlmann was also driven by something that seems common to many visitors, 'a desire to make a pilgrimage to Iceland in my capacity as an inhabitant of the north', and in this he was not disappointed, describing the higher reaches of Vatnajökull as 'like an ocean, not as in a storm, with foam-crested waves, but in calm weather, with long rolling ground swells disturbed by no ripple of wind'. Ahlmann's account of the successful expedition was published in a popular book which by 1938 had been translated into English as *Land of Ice and Fire* and included a technical appendix on the science: *Concerning Glaciers, and Vatna Jökull in Particular*. The book reinforces the value he placed on cooperation and working with local scientists and the communities as a whole. He encouraged Jón Eyþórsson and Sigurður Þórarinsson to continue his measurements and by the end of the decade they had a good understanding of the character of mass balance across the ice cap and had started working on Hofsjökull to the west, looking at the rate of losses on this different smaller ice cap. While camping on Vatnajökull Jón Eyþórsson mentioned that he had seen old documents concerning north Icelanders who had rights to fishing off the south coast and who were believed to have walked over the ice cap to get to their boats. Ahlmann felt that this would have been possible through a depression they were following across the ice south of Kverkfjöll: 'We could picture them to ourselves quite well, in their home-spun clothes, with long hose, tunic and hood, riding their shaggy little ponies across the glacier.' Their experiences tended to reinforce the view that Vatnajökull had probably been crossed quite regularly by farmers over the centuries. When Ahlmann died in 1974, a memorial published by the Geological Society of America described him as the Grand Old Man of Geology and Geography and a man of deeply founded democratic convictions, and his obituary in the *Geographical Journal* called him a 'somewhat legendary figure'.[5] This contrasts with a review of *Land of Ice and Fire* in the English journal *Nature* in 1938 by 'R.N.R.B.' which, while outlining the difficulties and successes of the expedition, could not help criticizing the language and managed to point out in a review of only three short paragraphs that Vatnajökull had been crossed in a double journey in 1932 by a Cambridge expedition and yet this was not mentioned at all, which reflected more on the condescending attitude of the reviewer than on the accomplishments laid out in the book.[6]

Ahlmann's involvement of Icelandic workers proved an important development in contrast to scientists and explorers from other countries, especially Germany and the UK, who tended to see the country as a no-man's land open and free for exploration. In 1939, five years before full independence, Iceland passed a resolution requiring foreign scientists to demonstrate collaboration with an Icelandic colleague before getting permission to carry out research. British university expeditions to Icelandic glaciers, especially from Cambridge in the 1930s, primarily saw exploration of novel regions as a goal in itself, with any scientific observations as an adjunct to provide respectability. As Auden noted in his 'Letter to Lord Byron', published in 1937 in *Letters from Iceland*, 'At Hvítarvatn and at Vatnajökull, Cambridge research goes on, I don't know how.' According to Ahlmann, German expeditions at the same time were considered by the Icelanders as tourists, and he was particularly concerned at the vulnerability of a newly independent country at a time of war with a danger of subservience to Britain, Germany or America.[7] The female German glaciologist Emmy Mercedes Todtmann (1888–1973) did, however, successfully return to Iceland on several occasions between 1930 and

1960, concentrating on studying a range of features around Vatnajökull.[8] Her last visit was in 1972 at the age of 84, and prior to this at the age of seventy she spent time at 4,000 m (13,125 ft) studying glacial features in the Andes in Venezuela. Some of the other expeditions from this time did produce useful observations and especially photographs that proved of scientific value. The publication of the visit to Vatnajökull in 1936 by Andrea de Pollitzer-Pollenghi, which explored the western part of the ice cap including areas crossed by Watts, was accompanied by a small number of excellent photographs of Grímsvötn, ice-surface features and their equipment.[9] They undertook some survey work but in opting for lightweight equipment, because of the need to go a long way in a short time, acknowledged the reduced accuracy of the results compared with a proper geodetic survey. Overall, though, the account still focuses on the exploration aspects rather than the science, and in the 'Practical Hints' section includes comments such as:

> Farmers do not generally go mountaineering . . . [they] ignore the use of skis . . . do not even wear common boots with a sewn sole, but [wear] the antique sheep-skin shoes . . . Alpine guides do not exist in Iceland.

At present not only is the extent and 'health' of Iceland's principal ice caps constantly monitored but there is a much greater understanding of how they move and react to changes in climate. Detailed observations have proved essential, due to the reliance of Iceland on hydroelectric power, as some of the main rivers used to generate the power have glacial origins. Hofsjökull in particular has been well instrumented as it feeds Þjórsá, Iceland's longest river, and has six hydroelectric power stations along its length. Recent mass balances have been negative, as glaciers throughout the country have been receding, but part of the ice cap did actually show a positive balance for 2017.[10] Even if Þórður Þorkelsson Vídalín did not accept the view of those living close to the ice of Vatnajökull that snow built up high on the glacier, transformed into ice and moved to lower ground, we now have a good understanding of glacier flow and how positive balance from the accumulation area of a glacier moves into the lower ablation zone. In Iceland the rates of velocity of such flow have been measured and, as in other glaciated areas, show quite a range of figures, with increased rates generally in the summer.

For Skaftafellsjökull, on the southern margin of Vatnajökull, estimates of these rates have been confirmed in a tragic manner. In 1953 two members of the British Nottingham University Vatnajökull

Hiking equipment from the 1953 Nottingham University Expedition found on Skaftafellsjökull in 2006 and now on display at Skaftafellsstofa.

Expedition, Ian Harrison and Tony Prosser, were lost in bad weather while carrying out research below the summit of Öræfajökull. Despite an extensive search they were never found. In 2006 remains of camping equipment, including a penknife, tent pegs, paraffin stove and tent poles, were discovered emerging from the ablation area of Skaftafellsjökull about half a kilometre (⅓ mi.) above the snout. As the main 'Ice Camp' that had been established at around 1,200 m (3,940 ft) lay outside the catchment of Skaftafellsjökull, it would appear that the equipment represented a camp established by the two when caught in bad weather. Comparison of the suggested location of such a camp, the time taken for the remains to travel down the glacier and estimates of ice velocity support this hypothesis, giving a rate of 0.4–0.5 m (c. 1½ ft) per day.[11] As their bodies have not yet been found it also seems likely that they were not in the camp when they died and were probably trying to escape to low ground or to the other better-equipped camp. The equipment is now rather poignantly on display at the Skaftafell visitor centre, Skaftafellsstofa.

There have been other recorded losses which have yet to be found that might prove similarly useful in understanding the rates and patterns of glacier movement. While searching for the source of the Skeiðarárjökull *jökulhlaup* in 1919, Hakon Wadell (1895–1962) was unable to collect heavy equipment left quite high on the ice cap west of Heinabergsjökull.[12] He also lost two horses and scientific material, none of which has yet been redis-covered. As the glaciologist Jack Ives said, quoting Ragnar Stefánsson of Skaftafell, 'the glacier always gives up what it takes,' so they should appear at some stage in the next few decades unless they have perhaps been lost in meltwater sediments. Hakon Wadell incurred the displeasure of Sigurður Þórarinsson for, with his fellow Swede Erik Ygberg, locating Grímsvötn and calling it Svíagígur, Swede's Crater. In a lecture to the British Geologi-cal Society at Cambridge in 1952 Sigurður Þórarinsson begged that respect should be given to the old Icelandic names and that new names should not be used.[13] This reflects the phenomenon of the importance of the linkage between place and identity in Iceland identified by Kirsten Hastrup,[14] 'words (including names) remain the most significant remains of the past,' and expanded upon by M. Jackson (see below) when Jackson says: 'When certain names of glaciers become more commonly used than others, invariably words, stories, and histories reduce or fade.' Giving names to locations in a foreign country also underlies a view of foreigners still seeing Iceland as a land to be explored and somehow belonging to those that 'discover' it. One of the biggest potential items that could have been used to trace ice velocities and movement patterns was a Douglas C-54 Skymaster cargo plane (registered name *Geysir*) that crashed on northwest Vatnajökull in September 1950 while travelling from Luxemburg to Reykjavík. It took two days for rescuers to reach the aircraft, where they discovered that the six-man crew had survived the crash and were supporting themselves on the cargo of expensive furs and Swiss chocolate. Regrettably the plane also survived relatively intact and was salvaged, eventually flying out of Iceland; a real opportunity lost to glaciology.[15]

One advantage of accumulating a wealth of data concerning rates of glacier advance and retreat since the 1930s has been the ability to identify glaciers that suddenly advance or surge. Once again it was Sveinn Pálsson who was the first to describe such glaciers, and post-surge the disrupted and chaotic nature of the ice surface provides good evidence of what has taken place. Sigurður Þórarinsson in the 1960s brought together a wide range of evidence that showed several glaciers had surged in the twentieth century despite the general pattern of glacier retreat, and it is now known that surges occur on all the major ice caps as well as at three glaciers in the Norðurlandsjöklar group. Some glaciers surge relatively regularly whereas others have a more irregular sequence of behaviour with gaps of between ten and a hundred years separating events. The shortest documented period is of ten years for Múlajökull on the southern side of Hofsjökull, while Síðujökull on southern Vatnajökull regularly surges every thirty years. Rates of glacier advance during a surge can be up to two to three orders of magnitude greater than rates during the non-surging periods. When Brúarjökull, a major northern outlet of Vatnajökull, surged in 1890 it advanced 10 km (6¼ mi.) in a matter of a few months. In 1875, having completed the glacial part of his Vatnajökull journey, Watts was surprised to find that 'a huge tongue of glacier at this point swept down to a distance of some ten miles beyond its northern limit,' almost certainly the result of a glacial surge and not surprising given that 76 per cent of the ice cap has outlets subject to surging. The causes of glacial surges have troubled glaciologists for decades.[16] Thoroddsen was able to rule out a volcanic cause from his observations and overall it is now accepted that

summer sun shines on it, a bit as if the countryside were blessed, and then its people would die rather than move away.

According to conversations recorded by Kirsten Halstrup, Snæfells-jökull, which had a reputation for magic, is believed to produce streams (*straumur*) of good and bad influence. To some extent the artworks of Katie Paterson, who sought to record the sounds of the glaciers Langjökull, Snæfellsjökull and Sólheimajökull on micro-groove ice records before playing them, thus causing the ice to melt and lose all the records they held, are a further appreciation of the perceived importance of the living presence of glaciers.[21]

Without needing to measure the position and extent of glaciers there is a very clear awareness of the way glacier margins have changed over the last eighty years that lies within individual and societal memories. At Hoffellsjökull, for instance, locals recall when it was possible to park by the glacier front to collect ice for use at Höfn in the fishing boats, but this has not been possible for many years as the glacier front has receded. Responses to glacier changes vary from sadness at losses to a feeling that it is all part of changes seen and experienced over the centuries by earlier generations and to which they adjusted and coped. However, placing that knowledge in the context of longer timescales of change, not just decades and centuries but millennia, is of immense importance if it means we are better able to communicate the likely abnormality of what is currently being seen; this provides the focus for the next chapter.

MOST UNIMAGINABLY STRANGE

CHAPTER 8

TILLITES, ICE SHEETS, GLACIER CHANGE AND PERMAFROST

'I SHALL NOT BE SURPRISED IF THIS ACCOUNT OF THE OCCURRENCE OF
GLACIAL DEPOSITS AND STRIATED ROCK SURFACES IN CONNECTION
WITH THE "PALAGONITE-FORMATION" OF ICELAND IS RECEIVED WITH
INCREDULITY.'[1]

'AÐ VERA EÐA VERA EKKI — JÖKULL.'
(TO BE OR NOT TO BE — A GLACIER.)[2]

Following the findings, by Louis Agassiz and others in the Alps, that
glaciers had been much more extensive in the past (and not only during
one Ice Age but over several different Ice Ages), a number of foreign
scientists, particularly the German Konrad Keilhack[3] and Carl Wilhelm
Paijkull, a geology professor from Uppsala whose book *A Summer in Iceland*
was translated into English in 1868, visited Iceland to look there for Ice Age
evidence.[4] They found striated rocks, erratic boulders and other features
similar to those used as key evidence for glacial theory in the Alps but
struggled to grasp their full meaning. Paijkull recognized the importance
of glacial erosion in the Icelandic landscape but the full length of time
available for repeated glaciations was still not commonly understood.
Similarly, Keilhack recognized 'old' strata that looked like glacial deposits,
but because they were covered by lavas he assumed they could not be
glacial as they would have been too old. It was not until the end of the
nineteenth century, with the survey by Þorvaldur Thoroddsen, that the full
extent of the glacial evidence began to be revealed, although he himself did
not believe in repeated glaciations – rather he assumed a single continuous
glacial period. It was Helgi Pjetursson, the first Icelander to receive a
PhD in geology, who, having travelled widely across Iceland in 1899 and
1900, came to the conclusion that not only had there been more than one
glaciation but earlier glaciations dated back to the Miocene, the warmer
period preceding the Pleistocene. In presenting his work in 1900 and
recognizing the controversial nature of what he was presenting, he stated: 'I

shall not be surprised if this account of the occurrence of glacial deposits and striated rock surfaces in connection with the "palagonite-formation" of Iceland is received with incredulity.'[5]

By examining sequences of lavas and the intervening sedimentary rocks in cliff sections across Iceland, Helgi Pjeturrson discovered what he termed palagonite-moraines but are now known as tillites. These are lithified glacial tills, the deposits left by glaciers and characterized by a mix of angular pebbles (technically known as clasts) of varying origin lying within a clayey matrix: sediment picked up and transported by glaciers, possibly over long distances, from their source. Þorvaldur Thoroddsen and others were often sceptical about such strata being glacial, especially in a country with extensive volcanic and fluvial activity, but the presence of striated pebbles and tillites lying on lava surfaces that were themselves striated proved to Pjetursson their glacial origin. Striated surfaces had been a central form of evidence for the proponents of glacial theory, nowhere more important than in the British Isles where, in the absence of present-day glaciers, striated surfaces, along with erratic boulders, were considered by Charles Darwin and Dean Buckland as incontrovertible evidence for the presence of a British ice cap in the past. Helgi Pjetursson's findings were supported by important geological figures. His work was reported in the prestigious *Journal of Geology* in 1900 by its founding editor T. C. Chamberlin (Thomas Chrowder Chamberlin, credited with the idea of multiple working hypotheses as a method for scientific exploration), who commented on how the results, although seeming incredible, were derived 'with apparent care and discrimination' and represented a very important contribution to the subject.[6] This must have delighted Helgi Pjetursson as he saw one of the essential aims of a scientist as being to 'practise one's ability to be observant'.

The identification of the palagonite-moraines was to prove Helgi Pjetursson's last major contribution to the subject, as he changed his name to Pjeturss in 1907 and turned his attention increasingly to what he called astrobiology, the study of psychic phenomena, believing that spirits contacted by telepathy were beings from other planets. His ideas are now termed cosmobiology and in 1990 the Helgi Pjeturss Institute was set up to further his ideas and carry out 'scientific research in the field of philosophy', reminiscent of the enlightenment of Edgar Mitchell from his time on the Moon following his experiences around Askja. Helgi Pjeturss was no doubt, like Björn Gunnlaugsson, aided in his thinking by spending many nights in tents undertaking his geological fieldwork in Iceland and Greenland, allowing him time to ponder on matters outside his geological research.[7]

In Iceland the presence of erratic stones had been recognized by locals, who termed them *Grettishaf* or 'Gretti's Lifts', and the Icelandic for an erratic is *Grettistak*. This refers to *Grettir's Saga*, or the *Saga of Grettir the Strong* (*Grettis saga Ásmundarsonar*), where, on being banished for killing Skeggi, Grettir demonstrated his strength by lifting a large rock:

> It was then that Grettir lifted a stone lying in the grass, which is still known as Grettishaf.[8] Many went afterwards to see this stone and were astounded that so young a man should have lifted such a mountain.

Subsequently he became the longest-living outlaw in Iceland, hiding in numerous caves and places around the country before ending his days on the island of Drangey off the northern coast. Baring-Gould visited the saga site near Geitlandsjökull and it is likely that his illustration of the glacier has the stone in the foreground. He finds the idea that Grettir could have ever lifted such a stone 'preposterous' and quickly returns to his interest in the exploration of Grettir's home, the 'mysterious' rich, grassy hidden valley of Þórisdalur. While there he could not resist a little excavation and paid a dollar to dig into a mound looking for Grettir's skull, but only found a large immoveable rock. Elsewhere in Europe, with growing interest in glacial theory and the idea of a widespread Ice Age, boulder-watching clubs were set up with the aim of searching for, identifying and preserving erratic boulders. In Britain these were concentrated in Lancashire, Yorkshire and Lincolnshire where, between 1870 and 1914, men and women regularly walked over areas looking for erratics and unfortunately occasionally collected them and removed them to parklands.[9] Similar groups also formed in France, looking to preserve the erratic record. In the Alps erratic boulders are named after the founding fathers of glacial theory such as Agassiz, de Charpentier and Venetz, and one of the most well-known erratics, the Pierre des Marmettes, is owned by the Swiss Academy of Sciences. Because of the ubiquity of glacial features in Iceland, glacial erratics do not have the same national status as those in mainland Europe, although some large boulders transported by *jökulhlaup* do have their own identity and history.

Confirmation of the idea of multiple glaciations dating back to before the Pleistocene had to wait for the development in the middle and late twentieth century of palaeomagnetic and radiometric dating of the lava sequences within which the tillites are found. Examination of cliff sections and boreholes across the country have now revealed direct evidence for at least 22 separate glacial/interglacial episodes, the oldest being found below a lava dated to 4.01 Ma. The pattern that is emerging is one of scattered highland glaciation from possibly as early as 7 Ma, with initial more substantial ice build-up between 4 and 3 Ma, principally in the form of glaciers developing on higher ground as individual ice masses. This continued between 3 and 2.5 Ma, into the earliest part of the Pleistocene, until after 2.4 Ma an inland ice sheet probably built up during cold periods, leading to the countrywide distribution of glacial deposits.[10] The sequence and the dates of the individual glacial episodes fit well with the offshore marine record, which is found in deep-sea sediments that represent changing water temperatures and the inputs from terrestrial glaciers. At the marine cliff site of Tjörnes just north of Húsavík, which has deposits dating back to 9.9 Ma, the period after 2.5 Ma has strata representing fourteen glaciations. The effects of repeated glaciation over such a long time can be seen in the deeply incised valleys of Iceland, which now expose extensive series of strata. During the earlier part of the Pleistocene, from 2.5 to 0.5 Ma, valleys 400 m (1,312 ft) deep were excavated at an average erosion rate of 10–20 cm/4–8 in. ka^{-1}; this accelerated to 50–175 cm/19¾–69 in. ka^{-1} as glaciation intensified after 0.5 Ma, deepening valleys by up to 1,000 m (3,280 ft). In the Holocene over the last 12,000 years the rate has dropped off to 5 cm/2 in. ka^{-1}, comparable with the period before 4 Ma when periods of glaciation only affected the highlands.[11] The impact of these cycles of glaciation on Iceland is therefore still very evident in the landscape we see today, producing some of the most stunning mountain landscapes such as that of Kirkjufell on the Snæfellsnes peninsula near Grundarfjörður, claimed to be the most photographed mountain in Iceland. It also reinforces the plea made by Pjetursson over a century ago:

Hitherto it has chiefly been volcanic Iceland which has received the attention of geologists. It is to be hoped that many will be able ere long to satisfy themselves by visiting our country that glacial Iceland is not less deserving of study.

At the time of the maximum extent of ice between 24 and 18 ka, known in the geological stratigraphy as the Weichselian,[12] Iceland was almost certainly entirely covered by an ice sheet up to 2,000 m (6,560 ft) thick that extended out onto the continental shelf. There is still some debate over whether a few peaks or nunataks survived above the ice or whether there were some protected ice-free enclaves around the coasts, but in essence there was a very large continuous Icelandic Ice Sheet (IIS) depressing the relatively thin crust. Climate then warmed and by around 14 ka almost 75 per cent of the ice sheet had disappeared as the warmer

Sabine Baring-Gould, 'Geitlandsjökull [part of Langjökull]', possibly with the insertion of Grettir's stone in the foreground, illustration in *Iceland: Its Scenes and Sagas* (1863).

ocean waters led to the removal of ice from the continental shelf. The rapid disintegration of the ice sheet produced responses in the Earth's crust. The weight of ice had depressed the crust, known as isostatic depression, and the reaction to the removal of the weight allowing the crust to rise, although rapid, was still delayed behind the flooding of the oceans from global melting of ice masses (known as eustatic change). It is possible to find abandoned shorelines up to 150 m (490 ft) above present sea level in western Iceland dating back to 15 ka as a result of this imbalance. With the continued break-up of the IIS and the rise in sea level as more ice was transferred globally from land into the oceans, sea level then fell to as much as 44 m (144 ft) below present levels by 10 ka as the land continued to rise, overtaking the eustatic effects.

Overall the end of the last glacial period and the early millennia of the Holocene was a time of considerable landscape change due to the local interplay of glacial disintegration and these isostatic and eustatic changes. This is indicated especially in the landscape of the southern lowlands, where melting ice and rising/falling sea level has produced a complex landscape. This complexity was added to by the additional pulse of volcanic activity that occurred as a result of the loss of the ice overburden and the release in pressure on the crust. It has been estimated that eruption rates immediately following deglaciation were between thirty and fifty times that experienced recently. The mix of processes is reflected in the landscape of the lowlands of southern Iceland, where there is a suite of glacial moraines marking one of the main limits of the IIS as it receded, the Búði complex.[13] To the uninitiated, picking out this feature within such a complicated landscape provides a real challenge, as it does even to Quaternary geologists. To earlier travellers the moraine would just have been a further series of low hills and mounds, but probably a welcome relief from the endless lava fields.

The glaciers of Iceland more or less reached their current positions and groupings by around 9,000 years ago, two millennia or so into the current interglacial, the Holocene. We now know that they have oscillated considerably since this time, and that there was a period in the Holocene rather grandly termed the Holocene Thermal Maximum (HTM) when mean temperatures were higher than during later millennia. So, the first question to tackle is whether ice caps ever fully disappeared during this warmer time. Determining the absence of an ice cap is not something that can easily be done in the field when the ice cap is now in place, unlike finding features formed by now-extinct ice. Local farmers and travellers found evidence for reduced glacier extent in the form of trees and turf emerging from the ice and transported by glacial rivers, and documentary evidence for former farms that were later overwhelmed by ice, but until the twentieth century the age and scale of these recessions could only have been guessed at. Over recent years two approaches have been used to assess the likelihood of ice cap survival. In the first of these, sediments are examined in lakes found within glaciated catchments. For Vatnajökull this has been from Lake Lögurinn, or Lagarfljót, eastern Iceland, which is fed by rivers originating from the Eyjabakkajökull outlet glacier, and for Langjökull it has been from Hvítárvatn, which lies immediately in front of the Norðurjökull outlet of the ice cap.[14] Lagarfljót is notable as the home of Lagarfljótsormur, or the Lagarfljót worm, the Icelandic equivalent of the Loch Ness Monster, which has been written about since 1345 and was filmed in 2012 with views divided as to what actually was represented. The video, by local farmer Hjörtur Kjerúlf, was considered to show a genuine monster by a committee from the local municipality Fljótsdalshérað, allowing him to collect a 500,000 ISKr prize. The monster is referred to on the Ortelius map as 'in hoc lacu est anguis insolitæ magnitudinis' (in the lake appears a large serpent) and has been seen on a number of occasions over the centuries, including several sightings considered reliable in the twentieth century.[15] Tales of monsters are found in other lakes in Iceland. Baring-Gould recorded meeting friends on his travels who had just missed a sighting of a *skrýmsli* or *skrímsli* by farmers at Skorradalsvatn in western Iceland, a 14-metre-long (46 ft) water monster with two humps.

Rivers that appear 'milky' were understood to have a glacial origin by Icelanders who lived close by and this information was passed on to

View across Lagarfljót with the inflowing river from Hengifoss.

fuel. Loss of land to ice from farms had wider effects on communities that were noted in annals of the time, such as cutting off access to mountain pastures or to limited areas of surviving woodland. It also impacted on rights to sections of shoreline to the south of the farms. The beaches of southern Iceland provided driftwood, which was essential for building as it comprised pines and firs from Canada and Siberia, much larger than the small birch trees that could be found locally. For the farmers south of Vatnajökull seals also provided a valuable resource, with annual seal hunts when time could be found away from farm duties. The glaciologist Jack Ives followed Ahlmann's tradition by spending extra time during the 1953 expedition on experiencing the farming life at Skaftafell with the farmer Ragnar Stefánsson. As part of this he went on the bi-annual seal hunt at Skaftafellsfjara, a challenging walk through soft sand and skua colonies. After a stew of pemmican, skua and seagull, the party proceeded to catch seals on the foreshore before they could retreat back into the sea. The sale of the pelts provided useful extra income and by carefully monitoring the size of the culls, past farmers had managed to retain the size of the population that provided them with income from pelts, meat and oil. This link between farms around the glaciers and rights to the shoreline ensured that any changes in the status of the farms as glaciers advanced would have been recorded, as would any effects on the rights of the surviving farms.

MOST UNIMAGINABLY STRANGE

Patterns of historical glacier change around the southern margins of Vatnajökull focusing on Skaftafell and Breiðamerkurjökull have been summarized by Jack Ives in his book *Skaftafell in Iceland: A Thousand Years of Change*. He based his summaries on his own research, earlier extensive work by Sigurður Þórarinsson[19] and more recent detailed studies that use a wide range of sources – from oral accounts and photographs, to techniques such as LiDAR (an acronym for light-detecting and ranging, the satellite-based measurement of a surface by sending a laser beam onto a surface and measuring its reflected character). This technique allows reconstruction of the three-dimensional geometry of the glaciers in the form of DEMs (Digital Elevation Models). During the time of settlement, it seems likely that Vatnajökull was still a large and significant ice cap but the margin lay some distance behind the current position, probably several kilometres inland. The region was well suited to the sort of farming familiar to the Norse settlers, with meadows and a variety of resources at hand. As climate went through colder periods within the general period of the Little Ice Age, broadly between 1250 and 1870, the ice advanced and as seen above some of the settlements disappeared beneath the ice. At Skaftafell the two glaciers Skaftafellsjökull and Svínafellsjökull had coalesced by 1708–9, blocking access to upland pastures, a feature also reported later in the century by Bjarni Pálsson, who described two large glaciers that were said to be always growing and were united. Along its western margin Breiðamerkurjökull had joined with Fjallsjökull by the time of the 1902–4 mapping by the Danish Geodetic Survey. The evidence in general points to glaciers having advanced close to their most extensive positions in the Holocene in the mid- to late eighteenth century and they must have remained similarly extended in the earlier part of the nineteenth century. Prior to this, Eggert Ólafsson and Bjarni Pálsson commented on the Breiðá river, which drains Breiðamerkurjökull, being the shortest river in Iceland at a Danish mile (equivalent to less than a British mile), and by 1836 a Frenchman, Paul Gaimard, claimed the distance had shrunk to 400 m (1,312 ft). Henderson in 1814 experienced the reality of advancing ice:

> Of its progress towards the sea, I was furnished with the utmost proof . . . I was surprised to find it traversing the track in the sand made by those who had travelled this way the previous year . . . and I again discovered a track, which had been made only eight days previous to my arrival, lost and swallowed up in ice.

Because Henderson was travelling during one of the coldest periods in the Little Ice Age, he had little experience of receding ice fronts but, as in the

Skaftafellsjökull in 2006 showing intense melting of dead ice in the glacier foreland area behind the late 19th-century outer moraine.

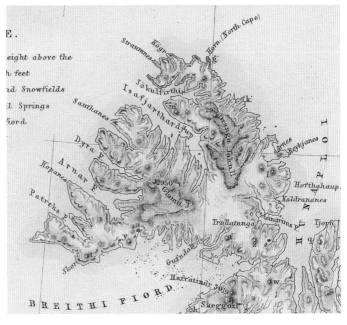

Detail of the Petermann map, based on Gunnlaugsson, showing the presence of both Glámujökull and Drangajökull.

Debris-covered glaciers can slow sufficiently to become covered with an increasing thickness of rock and sediment, transforming them into rock glaciers. The thick cover prevents further ice loss and the feature begins to take on a very characteristic shape, with transverse furrows across the surface and a steep frontal area. Although still active, rock glacier annual velocities are reduced to less than 1 m/3¼ ft a⁻¹, moving as much in a year as some glaciers can in a day, and they can remain remarkably similar in form and location for decades if not centuries. The differentiation between debris-covered glaciers and rock glaciers is usually taken as the former having an ablation area with around 50 per cent debris cover, itself more than 0.5 m (*c.* 1½ ft) thick and featuring longitudinal ridges, whereas a rock glacier has a cover between 40–70 per cent and a surface that has the very distinctive pattern of transverse ridges and hollows. In Iceland rock glaciers are mainly found at higher altitudes in the Norðurlandsjöklar group of glaciers on the Tröllaskagi peninsula. A recent inventory has identified 265 permafrost landforms across the Tröllaskagi peninsula, of which 178 are rock glaciers (118 active and 60 relict) and 87 are ice-cored moraines.[24] Within earth science rock glaciers have proved a topic of keen debate. Globally there has been significant disagreement between those who argue that some rock glaciers evolved from true glaciers through debris-covered glaciers to fully formed rock glaciers, and others, based principally in the European Alps, who argue that true rock glaciers are a feature of permafrost (permanently frozen ground, that is, ground that has remained below 0°C/32°F for two consecutive years) and do not include glacier ice.[25] A secondary issue is whether the existence of a rock glacier of glacial origin requires the presence of permafrost to survive.

Accepting the former definition for rock glaciers studied in northern Iceland, there is further uncertainty about just how long these features have taken to develop. At present there are three schools of thought: 1. The rock glaciers are of considerable age, having derived from the disintegration of ice at the beginning of the Holocene, and were reactivated during colder phases, particularly during the Little Ice Age. They are therefore of millennial longevity. 2. They formed during the Little Ice Age as glaciers expanded and became covered in debris by erosion of the surrounding cliff faces; they are thus centennial in age. 3. They formed as a result of neoglaciation and have persisted for several thousand years, conditions during the Little Ice Age enhancing their preservation; thus they are of millennial age but have not survived from the end of the last Ice Age. Because of their relative isolation they are not features that have had much attention from non-specialists despite their beauty and fascination

View of Sólheimajökull, an outlet of Mýrdalsjökull, the longest glacier in Iceland, which is now rapidly receding.

OVERLEAF
Brandi valley, Tröllaskagi, showing debris-covered glaciers of Bröndujökull and rock glaciers with extremely steep frontal slopes.

Tröllabrauð, Troll's Bread. Frost-shattered boulder on a glacier foreland in southern Iceland.

Þúfur on the valley side of Skíðadalur in Tröllaskagi.

negotiate a field of *þúfur*, hummocks varying in size from footballs to sofa bolsters, to reach his new home,[30] and the Reverend Frederick Metcalfe talks of fields looking like 'aggradations of anthills' or churchyards, as did Mrs Alec Tweedie when negotiating them on horseback for the first time. *Þúfur* are typically between 50 and 150 cm (19¾–59 in.) in diameter, varying in height from <10 cm to 1 m (4 in.–3¼ ft), and are particularly noticeable in abandoned fields within drained lowlands in southern Iceland, although they occur widely throughout the country. They are thought to form due to deep seasonal freezing, uneven snow distribution across the surface and impeded drainage, especially where the main parent material is dominated by volcanic wind-blown 'loess'. Deep winter freezing aids the formation of cracks and the building up of the mounds between the cracks, exposing the mound surface to freezing; hence it would appear that the pattern of *þúfur* formation over time mirrors the general pattern of climate variability seen over the Holocene with development in particular during the cooler periods. Grazing pressure can, though, exacerbate their formation, with sheep and especially horses widening the cracks. Deep snow cover provides a degree of protection for the surface and will reduce the tendency of mounds to form. Where tephrochronology has been used to date *þúfur*, it would appear that they reformed with neoglaciation around 4,500–4,000 years ago, and again around 2,600 years ago, although with more stable soils at times between. During the Little Ice Age there were at least two pulses of formation.[31]

What is apparent to anyone revisiting Iceland is just how quickly these hummocks form in fields that are no longer drained. Predicting the future of possible permafrost areas in Iceland is not just a matter of increased temperatures leading to less permafrost, as snow can protect the surface, hindering permafrost development. Between 2010 and 2016 windy locations on Tröllaskagi were found to have permafrost developing at altitudes up to 400 m (1,312 ft) lower than at sites which were less windy. Thus a future climate with less snow but stronger winds could mean little reduction in the areas prone to permafrost development. Reduced permafrost in mountainous areas would, though, lead to enhanced slope instability and the possibility of a more hazardous upland environment. However, given the location and sporadic nature of permafrost in Iceland, the country should be safe from the release of dangerous viruses and bacteria such as has recently been reported from Siberia,[32] presented even in 'respectable' media such as NPR (U.S. National Public Radio) as *The Zombie Diseases of Climate Change*.[33]

Skeiðarársandur with embankments to channel glacier floods, taken in 2010 after the major 1996 *jökulhlaup*.

Jökulhlaup have been described in annals and travellers' accounts over the centuries, notably when having an impact on surrounding settlements. In 1362 it was the *jökulhlaup* associated with the eruption of Öræfajökull that devastated the local surrounding farms, discussed in detail in the next chapter, a process repeated in 1727.[7] When Henderson visited in 1814 he was shown the area affected in 1362 and included in his account a first English translation of a letter from Jón Þorláksson to Secretary Olavius describing the 1727 eruption and *jökulhlaup*. The eruption started on 7 August 1727 when Jón Þorláksson was delivering his sermon at the Sunday service in his church at Sandfell on the slopes below Öræfajökull. The following day he saw 'several eruptions of water' gushing out of the glacier before 'the ice-mountain itself ran down into the plain, just like melted metal poured out of a crucible'. The flood removed all the remaining pastureland and as it approached his house Jón Þorláksson removed his family, belongings and church utensils to a high rock, where he pitched a tent and watched the remainder of the eruption. Some of the details of the accounts of floods such as that of Jón Þorláksson were treated with scepticism within scientific circles, but as the geomorphologist Andy Russell observed following the 1996 flood on Skeiðarársandur:

> Observations of recent jökulhlaups have brought to life descriptions of the historical jökulhlaups within the Icelandic literature and have reaffirmed our faith in the accuracy of historical observations made by the inhabitants of areas subjected to some of the world's largest contemporary floods.[8]

While the immediate flooding in a *jökulhlaup* may prove fatal it was also their geomorphological legacy that could be dangerous. The deposition of sands with high water content meant that areas of quicksand would be scattered across the formerly flooded areas. These were marked by warning signs after the 1996 flood, although from the evidence of footprints in the soft, sandy surface, they did not put off inquisitive visitors. In 1861, following a Skeiðarárhlaup, a *jökulhlaup* from Skeiðarárjökull, Jón Einarsson, from one of the local Skaftafell families, drowned in a kettle hole formed by a melting ice block left by the flood.[9]

Apart from the district adjacent to the southern margins of Öræfajökull, the other settled area affected by recurring *jökulhlaup* has been around the southern margins of Mýrdalsjökull, which was affected by eruptions of Katla and the associated *Kötluhlaup* (a term for all floods having their origins in eruptions of Katla). Henderson recounts these in some detail showing how *Kötluhlaup* repeatedly devastated communities, especially when directed through narrow valleys rather than flowing across the broader area of Mýrdalssandur, yet even as the floods reached the low-lying area of Álftaver to the east they were still of sufficient magnitude to cause serious damage to pasture and settlements. As Katla is one of the most active volcanic systems over the Holocene, with at least twenty eruptions since settlement, it is highly likely that each eruption would have had associated floods. In 1660 an outburst through Höfðabrekka was described with the church seen disappearing in the ice flood, as quoted at the beginning of this chapter. During the 1727 eruption a flood lasting three days 'carried . . . such amazing quantities of ice, stones, earth and sand, that the sea was filled with them, to the distance of three miles from the shore', and in 1755, during what was probably the largest historical flood in Iceland, estimated at up to 400,000 cumecs, 'Masses of ice, resembling small mountains in size, pushed one another forward, and bore vast pieces of solid rock on their surface.'[10]

From the nature of deposits produced by *Kötluhlaup* it has been possible to estimate past magnitudes and the most recent 1918 eruption provided observations against which these estimates could be evaluated. Over 400 m (1,312 ft) of ice was melted within four hours and once through the ice the eruption produced an ash cloud that reached an altitude of 14 km (8¾ mi.). A trough 2 km long, 500 m wide and 200 m deep (1¼ mi., 1,640 ft, 656 ft) was cut through the glacier leading on to the *sandur* where 8 m (26¼ ft) of sediment were deposited, including massive ice blocks that can be seen from photographs at the time. Large boulders – such as Kötluklettur, a block estimated to weigh 1,000 tons that had been transported 15 km (9¼ mi.) – that sit isolated on *sandur* surfaces are a

Jökulhlaup-transported boulders in Markarfljót.

JÖKULHLAUP, WATERFALLS AND CANYONS

characteristic feature of *jökulhlaup*-derived outwash spreads. Floods of
this type can achieve the status of what are known as hyperconcentrated
flows, where the sediment is so highly concentrated that particles move
together, suspending and effectively carrying the coarser material. Flood
discharges for Katla have been estimated at various discharges up to the
400,000 cumecs of the 1755 event, with peak discharges lasting a matter
of hours. In order to give some meaning to these figures it is usual to
compare them with the measurements for rivers around the world. For
Great Britain the average discharge of the Thames in London is less than
70 cumecs, and in Germany the discharge of the Rhine is around 3,000
cumecs.[11] The most powerful global river in terms of discharge is the
Amazon at around 175,000 cumecs. For the purposes of hazard planning
around Katla and Öræfajökull, models are run with a discharge of 300,000
cumecs, hopefully a figure that will be at the high end for any future
events.[12] These measurements put into some perspective the immense
power of the *jökulhlaup* that have affected Iceland in the past and which are
likely to continue in the future. Estimates of some prehistoric *jökulhlaup*
have suggested figures in excess of 1 million cumecs and the magnitude of
such events is manifest in a landscape cut by canyons exhibiting waterfalls
of international recognition and with extensive spreads of sediment that
form the large *sandur* plains.

To the east of Mýrdalsjökull *Kötluhlaup* deposits cover around 500 km² (190 mi.²), with the patterns of flow being diverted around the small farming area of Álftaver by a combination of a branch of the Eldgjá lava flow and an earlier moraine, although the increasing height of the *sandur* over the years has caused floods to overtop the lavas in places. Behind the moraine there was originally a lake noted for the hunting of swans, hence the name – *álft* is swan, the whooper swan, in Icelandic. The swans were a focus for the original settler at Álftaver, Molda-Gnúpur, but with the Eldgjá eruption of 934–40 CE he had to evacuate to Höfðabrekka and this area is now marked by a series of rootless cones formed during the eruption, part of the Katla Geopark. Floods from Katla have gradually built out the coastline to the south, such that before the 1660 eruption boats could be sheltered in a cave below Vík, but since this time the distance to the sea has been too great to use this protection.

As the Icelandic ice sheet disintegrated at the end of the last glacial period, particularly in the millennia covering the eventual transition to the Holocene between 12,500 and 9,300 years ago, a combination of large volumes of meltwater, rising land and increased volcanic activity is assumed to have generated a series of intense *jökulhlaup* along the major glacial rivers.[13] These cut down rapidly into the bedrock, especially through beds of more friable hyaloclastites. One of the highlights of the Golden Circle tourist route is Gullfoss, which lies east of Geysir on the Hvítá. It lies at the head of the Gullfossgljúfur canyon, which is between 40 and 70 m deep, 3 km long (130–230 ft, 1¾ mi.) and cut through alternating

Holocene and probable earlier *jökulhlaup* routeways across Iceland identified by Brigitte Van Vliet-Lanoë and colleagues (published in *International Journal of Earth Sciences*, 109 (2020).

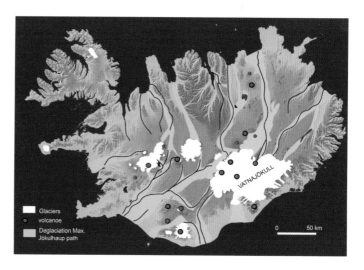

JÖKULHLAUP, WATERFALLS AND CANYONS

Gullfoss in winter looking downstream.

basalts, breccias and tuffs. During deglaciation it is believed that a large ice-dammed lake developed, Kjölur, which drained into the Hvítá river, helping to excavate the canyon with floods estimated at up to 300,000 cumecs.[14] For the Holocene as a whole a retreat rate of 30 cm/12 in. a^{-1} has been suggested for this valley, allowing the river to cut down and back, especially by undercutting the harder lavas, causing blocks to break off. Gullfoss has played an important role in conservation in Iceland as it was the scene of one of the earliest fights to protect rivers from development for energy generation. At the end of the nineteenth century an Englishman attempted to buy the falls from the owner of the farm at Brattholt. His

MOST UNIMAGINABLY STRANGE

offer was declined but subsequently a powerful Icelandic group took over the waterfalls with the aim of harnessing the river for power generation. The farmer's daughter Sigríður Tómasdóttir (1871–1957), who acted as a guide to visitors, took up the fight to prevent this from happening. She was supported by a lawyer, Sveinn Björnsson, who later became the first president of Iceland, and she famously walked the 130 km (80 mi.) barefoot to Reykjavík to make her case, as well as threatening to throw herself off the falls if they were not protected. Although she lost the court case against the developers, the government bought Gullfoss and agreed to protect it for the future, albeit that permanent conservation was only achieved in 1979, over twenty years after her death. The main problem at Gullfoss now is working out how to cope with its popularity and ever-increasing visitor numbers. The victory of Sigríður Tómasdóttir is often taken as representing the birth of the environmental movement in Iceland, eventually leading to the establishment of a series of national parks.[15]

Although protected, Gullfoss is not part of a national park. The belated creation of national parks in comparison with other countries possibly reflects an initial feeling that because the country was so isolated and rarely visited, many of the areas likely to be designated did not need formal protection. But as pressures have mounted from economic needs, especially hydroelectric power generation, and from tourism, potentially threatened land has gradually attracted legal protection. The largest park centred on the Vatnajökull ice cap, Vatnajökulsþjóðgarður, was established in the late 1960s, the exact date remaining rather uncertain depending on just how the necessary documentation is interpreted. The area covered has gradually grown from the immediate area of Vatnajökull behind Skaftafell in 1967 to the complete area of the ice cap and much of its glacier forelands, also embracing parts of Ódáðahraun including Askja and Herðubreið to the north and detached areas around Dettifoss and Jökulsá á Fjöllum. Þrándarjökull and Hofsjökull, glaciers to the east of Vatnajökull, are not within the current boundary. Protection outside national parks owed much to the efforts of Sigurður Þórarinsson in the decades after the Second World War. He campaigned tirelessly to conserve landscapes characteristic of Iceland's geological heritage such as the pseudocraters around Mývatn, while at the same time playing an important role in the designation of Vatnajökulsþóðgarður. Apart from Þingvellir and Vatnajökull, Snæfellsjökull was accorded national park status in 2001. There is now a move to create a single national park covering all the Central Highlands, or Hálendið, covering 40,000 km² (15,400 mi.²), around 40 per cent of Iceland, although there is some opposition from energy companies and from local authorities with land that would fall within the designated area.[16]

JÖKULHLAUP, WATERFALLS AND CANYONS

Gullfoss in 1934, photographed by Willem van de Poll.

> One of the most wicked-looking places one could ever see, a great
> scar in the plain . . . cliffs of black, sinister-looking basaltic rock,
> and at the bottom . . . a rushing, mad, angry, clay-coloured thick,
> muddy river, hurrying, crashing through the canyon, and awaking
> howling echoes from the perpendicular cliffs.

The discharge of the Jökulsá á Dal has now been reduced by half by the
Kárahnjúkar dam constructed upstream as part of the Fljótsdalur hydropower plant, which was built to provide power for the Fjarðaál aluminium
smelter to the east in Reyðarfjörður, although it still flows with a discharge
of almost 100 cumecs. It is perhaps noteworthy that Pike Ward made
no comment about the stunning basalt cliffs along the river, although
his interest lay more in getting from A to B in the safest and driest way
possible rather than in the beauties of the immediate landscape.

Most ice-dammed glacier lakes that provided sources for *jökulhlaup*
were to be found in isolated areas and rarely described but the first drawing
of an Icelandic glacier included reference to just such a lake. In 1704 Árni
Magnússon described how Sólheimajökull had advanced and created
a barrier to meltwater travelling down Jökulsárgil from Mýrdalsjökull.
Between floods the water travelled in a tunnel below the glacier until this
became blocked, which created a lake that eventually forced the glacier
to lift and slide, allowing the water to escape catastrophically. Floods
from Sólheimajökull play a part in a well-known Icelandic story found in
Landnámabók featuring two settlers who were also sorcerers, Loðmundur
at Sólheimar and Þrasi Þórólfsson, who lived to the east in the Skógar area.
Þrasi saw a large flood coming and moved the river nearer to Sólheimar.
In response Loðmundur told his slave to put a large stick in the water to
which he attached a ring to move the river back towards Þrasi Þórólfsson.
After much toing and froing they found a canyon for the river that avoided
both their lands and settled their differences. The 'mythical' flood is linked
to an eruption under Mýrdalsjökull and thus the underlying theme is
likely to be related to a real event. Þrasi Þórólfsson was also recorded in
folklore as having hidden a chest of gold and treasure below the Skógafoss
waterfall but unfortunately this has never been found, despite further
stories of occasional sightings in the water.

Apart from being important tourist attractions waterfalls have
provided a recurring theme for the artist Ólafur Elíasson. In 1996 he produced a set of fifty photographs of different Icelandic waterfalls tinted in a
range of colours to highlight their individuality, producing what one critic
suggested were images reminiscent of old postcards from exotic locations.
He also installed 'waterfalls' as part of exhibitions in a number of locations

had proved a very real challenge to anyone seeking to spend time in its empty landscape. Þorvaldur Thoroddsen identified the critical issue as a lack of grass to support the horses used by earlier 'explorers', with the large 1840 expedition led by T. C. Schythe failing for 'want of grass . . . Nearly all the horses died, whilst the members of the expedition barely struggled back to the inhabited districts alive.'

Around 22,000 km^2 (8,500 mi.^{2}) of Iceland is classified as sandy desert, a fifth of the country, mostly composed of basaltic volcanic glass derived directly from volcanic ash or reworked by wind and water in the extensive glacifluvial *sandur* plains around the ice caps. Big explosive eruptions such as Veiðivötn 1477 would renew the ash cover close to the fissure, as well as changing the landscape by producing new volcanic landforms. These desert areas are a focus for impressive dust storms that can move large volumes of sand, usually measured in kg m^{-1}, the amount of sand that can be moved over 1 m (3¼ ft) in distance over a set time. The biggest Icelandic source, defined as a hotspot or a confined plume area, is Dyngjusandur along the northern edge of Vatnajökull; here the combination of a rain shadow and extensive supplies of sand deposited by floods from the glacifluvial rivers draining the ice cap provide excellent conditions for suspension of dust by the strong katabatic winds blowing off the ice. Major events activating 1–5 kg m^{-2} of sediment have been measured, comparable to those experienced in the Bodélé depression in Chad, the planet's largest single dust source.[3] Fortunately the 30 km^2 (11½ mi.^{2}) of Dyngjusandur is tiny compared to the 200,000 km^2 (77,220 mi.^{2}) of the Bodélé Depression, which is twice the size of Iceland. Overall around 30–40 million tons of dust are moved in Iceland annually and most is redeposited, leading to high rates of soil accretion in some areas. Plumes have been shown by satellite imagery to travel over 500 km (311 mi.), reaching Ireland 1,300 km (808 mi.) distant and recently traced by geochemical analyses to Belgrade in the Balkans.[4] Iceland's contribution to global dust emissions is an order of magnitude greater than the Arctic as a whole and is similar to that found in warm deserts, contributing to the mineral enrichment of the surrounding North Atlantic. During dust events Reykjavík can experience air pollution levels that exceed WHO guidelines (Reykjavík Haze), and for locations within 30 km (18½ mi.) of the source these can be up to a hundred times the recommended limits, equivalent to figures from heavily industrialized areas – not something that sits well with an image of a clean, natural environment as often portrayed in adverts. Mrs Tweedie encountered a dust storm on her ride to Geysir and provides quite a vivid description of an 'extraordinary yellow haze' rolling towards them like a London fog from the interior, half-choking the party and turning their hair yellow-grey.

Volcanic desert area caused by the Veiðivötn 1477 eruption with hyaloclastite ridges in the background.

Although the dry desert areas of the Central Highlands owe their origin to a combination of their geology and location, and have probably remained remarkably similar over the Holocene timescale, constantly being sustained by new eruptions, other deserts are of more recent origin, either through the impact of volcanic activity or of humans. Öræfi, literally the wasteland or desert, the subject of Ófeigur Sigurðsson's 2015 Icelandic Literature Prize-winning book, a book he considered to be a private joke and did not expect to get published (it includes, for instance, eighteen pages of Annals extracts concentrating on suicides between 1400 and 1880), only derived its name following the 1362 eruption of Öræfajökull that devastated the area. Prior to the eruption the region was known as Litlahérað and, although isolated between the vast *sandur* plains in front of Skeiðarárjökull to the west and Breiðamerkurjökull to the east, it was one of the richest farming areas in the country. Mid-fourteenth-century records showed forty farm estates, twelve of which were owned by the four churches within the area, churches whose property in terms of livestock, missals and sacraments showed them to be as wealthy as any in Iceland. The 1362 eruption led to the deposition of 2 km³ (½ mi.³) of tephra over the immediate area and was accompanied by a series of six major *jökulhlaup*. No eyewitness accounts survived but a combination of later records and oral histories confirmed the total devastation and abandonment of this once-rich land, with resettlement not taking place until the end of the century and on a much smaller scale, comprising only eight farms, two still owned by the church. Reconstruction of the nature of this major event in Icelandic history owes much to the painstaking research of Sigurður Þórarinsson, who in 1962 combined his innovative tephrochronological expertise with a synthesis of the documentary, oral and physical evidence to reconstruct the nature of the eruption and associated floods, putting forward the view that despite the undoubted catastrophic impact of the flooding it was the widespread fall of tephra that impacted most on settlements and farmland, leading to the 'desertification' of the region. As Sigurður Þórarinsson summarized:

> When it rose again it no longer bore the name *Hérad*, not even *Litlahérad*. It was called Öræfi. This singularly paradoxical name of a human habitation, Öræfi, wilderness, anæcumen is the initial word in a new story, and at the same time the last word in the story of *Hérad milli Sanda*.[5]

This is a further example of Icelandic scientific writing that transcends the dry description of data and events and offers an emotive and existential context for the research. It also represents a recurrent theme among Icelandic and Scandinavian researchers in particular, a realization that insight does not always come from a deep, but perhaps too narrow understanding. Sigurður Þórarinsson, despite his scientific background, argued that his interest in volcanoes and glaciers came from seeing their importance as the milieu within which Iceland's people lived, not from just any intrinsic fascination. He made this very clear in a recorded interview he gave at the University of Lund in Sweden in 1979 (available on YouTube), and also took care to recognize the value of the breadth of his early home-based education. He was grateful to have learned Latin and especially some Greek so that when it came to looking for a word for volcanic ash to add to magma and lava, both with Greek roots, Aristotle's tephra was a logical choice.

The idea of catastrophic devastation of a previously productive landscape, either by flood or more likely by volcanic activity, had been invoked in the 1890s by Daniel Bruun excavating farm sites in Þjórsárdalur near Hekla, especially that at Stöng where he found farmsteads buried by pumice derived from the volcano. He coined the term 'Pompeii of the North' to describe his findings of a farm laid waste by ashfall in a single catastrophic event. Subsequent excavations in the summer of 1939 by a team of Nordic archaeologists concentrated on six farm sites and Sigurður Þórarinsson suggested, on the basis of the tephrochronology, that the culprit for abandonment could have been the well-documented Hekla eruption of 1104. This view generally held sway until later excavations and tephra analyses, carried out in an academic context more critical of simple single cause-and-effect explanations, suggested a less cataclysmic downfall for farms in the area, with greater resilience of people and ecosystems to such eruptions.[6] Whatever the effects, the term 'Pompeii of the North' now lives on elsewhere in Iceland – as, for example, applied to excavations on the island of Heimæy in Vestmannæyjar, where houses covered by ashfall during the 1973 eruption have been revealed and form the core of the Eldheimar Museum.[7] The details of the excavated farm at Stöng were used as the basis for the reconstruction of a Commonwealth-period farmstead nearby at Þóðveldisbærinn, which was used as another Icelandic *Game of Thrones* location.

Prior to the settlement of Iceland, desert areas were restricted to the Central Highlands and *sandur* plains, but with colonization by the Norse settlers, bringing with them sheep, cattle, pigs and horses, especially the sheep which roamed widely around the farms and over the mountain pastures, soil erosion accelerated. Once breached, the surface of the thick but relatively fragile soils built up by volcanic sands over millennia

Church at Hof below Öræfajökull. Built in 1884, it is one of the few remaining with a turf roof similar in form to those of earlier churches.

OVERLEAF
Turf walls from Glaumbær, Varmahlíð, northern Iceland.

Receding face of eroded soil in southern Iceland exposing bedrock.

proved highly susceptible to wind erosion.[8] Protected in part originally by woodland, as reflected in the frequently quoted description of Iceland at the time of settlement made later in the twelfth century by Ari Þorgilsson hinn fróði, Ari the Wise, in *Íslendingabók* – '*Í þann tíð var Ísland viði vaxit á milli fjalls ok fjöru*' (At that time Iceland was covered in trees from the shores to the mountains) – the loss of woodland accelerated soil erosion, but more especially it was the pressure of animals that eventually led to an estimated loss of 70 per cent of the country's soil. Now only 4,000 km² (1,540 mi.²) of the land surface of Iceland is considered free from erosion or subject to only minimal loss.[9] Once the soil is exposed, rates of deflation, the removal of soil by wind, can be very high with up to 40 cm (15¾ in.) potentially lost in a season. One of the most distinctive features of the Icelandic landscape, found across almost a fifth of the country, is the *rofabarð* (plural *rofabörð*, see Chapter Three), the eroded remnants of former soils seen as small cliffs or escarpments between 20 and 300 cm (8–118 in.) in height and in extreme cases forming islands protruding from a largely eroded bare surface. The exposed cliffs retreat rapidly due to a combination of wind action, water in the wetter western areas and freezing and thawing during the winter. Stabilization is not helped by sheep choosing to use them for shelter and further destroying the friable soil. Historical rates of erosion have been determined by analysing the density of breaks in the vegetation as the surface has been broken up, the rapidity of retreat along the breaks and the depth of the soil profiles affected. Amounts of loss reduced in later periods, but as deeper profiles are now being affected the long-term effect on the landscape is still significant. It has been estimated that current rates of soil

MOST UNIMAGINABLY STRANGE

loss reach 230 ha/570 acres a⁻¹ for the country as a whole, broadly equivalent to 230 football pitches or, perhaps more appropriate in an Icelandic context, almost 3,000 handball courts.

Concern over the extreme losses of soil led to the formation of the Soil Reclamation Office in 1907, later renamed as the Icelandic Soil Conservation Service (ISCS, Landgræðslan),[10] and considerable efforts to protect uneroded land and stabilize eroding soils have taken place over the century or so since its formation. Given the obvious role humans have played in destroying the essential basis for plant growth in the country there is something of a moral imperative to make progress in this area. With independence in 1944, a Land Restoration Fund, Landgræðslusjóður, was set up to reflect a national desire to foster revegetation, followed in 1974 by The Nation's Gift, or Þjóðargjöfin, a fund that also supported reforestation. In an early move to involve everyone in the activities in the early part of the millennium the Retailer's Association agreed to introduce a plastic bag fund, ahead of many other European countries, which charged shoppers for use of the bags, with the proceeds funding conservation schemes. By fencing off land it has been possible to prevent further erosion and the contrast in plant diversity and luxuriance seen either side of fenced-off parcels is striking. Reseeding with lyme grass, *Leymus arenarius*, mimicking its use to stabilize sand dunes in coastal regions across many other countries, has proved reasonably successful and can be seen on the sandy areas of the *sandur* plains along the south coast. There have also been efforts to protect remnants of birch woodland and expand the area under trees.

In 1908, a year after the formation of the Soil Reclamation Office, the Icelandic Forest Service (IFS), Skógræktin,[11] was formed to oversee protection and afforestation. The first organized planting had taken place in 1899 at Þingvellir when three Danes, two forestry experts and a merchant marine captain, Carl H. Ryder, planted the Pine Stand. Impetus towards reforestation was given by the first prime minister of Iceland, Hannes Hafstein, who was in office when the Forest Service was set up and in 1900 wrote a poem that included the lines: 'The time will come when tattered land regrows, Culture blossoms in the shade of woodland glades.'[12] Despite his efforts relatively little was achieved but in the 1980s Vigdís Finnbogadóttir made tree planting a feature of her presidency, and on official visits across the country planted three birch trees, 'One for the daughter, one for the son, and one for the future of Iceland.' In early 2020 the National Church in Iceland announced plans for a baptism forest, offering to plant a tree for every child baptized in Iceland.[13] Natural woodland in Iceland has always been very limited in species, dominated by birch, *Betula pubescens*, with limited contributions from rowan, *Sorbus aucuparia*, and the much rarer aspen, *Populus tremula*. Only in limited areas would trees gain heights of up to 15 m (49 ft), hence the hoary old joke: 'What do you do if you get lost in an Icelandic forest? – Stand up!' In 1900 Douglas Scott reported that the tallest tree in Iceland was a solitary rowan in Þórsmörk that reached 9 m (30 ft).

Protection of the extant birch woodlands has, though, been successful, with National Forests spread across the country, and there is now some expansion thanks to enclosures, even to the extent of harvesting birch logs for fuel and their use in pizza ovens! From 1976 there has been an attempt to recolonize with birch by establishing protected islands allowing the indigenous trees to flourish. Birch is naturally colonizing an area 35 km² (13½ mi.²) in extent on Skeiðarársandur, a feature first noted in 1996 and now being mapped by drone. Recent genetic analysis of the trees shows that they are from the Bæjarstaðarskógur woodland that covers over 20 ha (49½ ac) around Skaftafell.[14] With the growing realization that the climate of Iceland was not inimical to tree growth, planting of non-native species has increased and now Siberian larch (*Larix sukaczewii*), Sitka spruce (*Picea sitchensis*), lodgepole pine (*Pinus contorta*) and black cotton-wood (*Populus trichocarpa*) can be seen. Seedlings of black cottonwood were sent to Iceland from Alaska by Vigfús Jakobsson during the winter of 1943–4 but Hákon Bjarnason as Head of IFS deferred giving it an Icelandic name until it showed it could survive. In 1945 it was named as *Alaskaösp*, Alaskan poplar.[15] Around 130 different tree species have been planted to see how they fared in the Icelandic environment, with fourteen being considered suitable. While making economic sense, the presence of exotic species in a country which markets itself to tourists as natural and very distinctively Icelandic can create tension among conservationists, although public opinion appears not to be against exotic trees so long as they do not grow in extensive plantations. The Icelandic rescue organization ICE-SAR has also contributed to current attempts at reforestation by taking advantage of the country's predilection for fireworks that are sold through their outlets; over 600 tons are let off annually. Seedlings to be planted by the Forestry Service have been made available to sell with the fireworks under the slogan *Skjótum rótum* ('Put Down Roots') and in 2018 13,000 were planted.[16] ICE-SAR have committed to continue the practice until 2023.

More problematic has been the expansion of Nootka lupin, *Lupinus nootkatensis*, which was introduced from Alaska in 1945 to stabilize bare surfaces and prevent soil erosion, a species termed a 'Legume Ecosystem Engineer' by one researcher. It is now ubiquitous in Iceland and covers large areas of newly deglaciated terrain, *sandur* plains and river floodplains. Its striking blue flower can be seen from a distance and is commonly used as foreground for landscape panoramas found on postcards (there are now online guides for tourists showing where to see them), yet it is not only an alien plant but not of European origin, unlike the overwhelming majority of the native flora, and its introduction has initiated a very lively debate in Iceland spilling into broader issues of nationalism and perceived xenophobia.[17] It was the head of the IFS at the time, Hákon Bjarnason, who brought two spoonfuls of seed from College Fjord in Alaska to accelerate the process of testing the suitability of the lupin as a plant to stabilize bare soils. By 1955 its spread at Þverárdalur, where it was introduced, was considered a 'remarkable story' and for the next few decades it was seen as a very positive development, with aerial seeding first taking place in 1963. By the early 1990s people opposed to the introduction of exotic conifers, including many ecologists, broadened their fight to include the Nootka lupin and a polarized debate developed between them and supporters of the lupin who, along with Hákon Bjarnason, saw a benefit in widening species diversity and encouraging soil development.

In response to this increasingly vitriolic debate research has been undertaken to evaluate the impact of this plant on ecosystems,[18] research which confirmed its benefits for colonizing bare ground but also pointed to how it outcompeted dwarf shrub heath, an essential component of the native vegetation and a breeding ground for the culturally significant golden plover or *Lóa*, *Pluvialis apricaria*. The first appearance of the golden plover is eagerly awaited as the first sign of spring, its nickname of *vorboðinn ljúfi* usually translated as 'the sweet herald of spring',[19] and folklore has it that it was not created by God on the fifth day but later by Jesus when as a child he was making clay statues of birds. Successful colonization of riverplains by the lupin has also begun to have an effect on the breeding of the whimbrel, *Numenius phaeopus*, another bird for which Iceland is an important breeding ground, responsible for 10 per cent of the global breeding population.[20] As a result of the threats to ecosystem diversity attempts were made to restrict expansion of the lupin by mowing and also by spraying with Roundup, the Monsanto glyphosate-based herbicide – not a popular move with a lot of people for obvious reasons. The debate continues and with the advent of social media has moved from the pages of the newspapers to Facebook, with the Lupin Killers Group and Vinir lúpínunnar, the Friends of the Lupin group, attracting support into the thousands, the latter currently having over 3,000 supporters. Arising from this focus on invasive species and the non-native versus native clash, 2011 saw what Karl Benediktsson has described as the legal right to award Icelandic citizenship to plants being given to a group of experts as part of new nature conservation legislation. Any plant not on their list would be considered an alien.

Birch trees recolonizing the *sandur* plain at Skaftafell.

Prior to the 2011 ruling the 1999 Nature Conservation Act had provided the first legal definition of native species – they had to be included in the third edition of *Flóra Íslands,* published in 1948.[21] This was first produced in 1901 by Stefán Stefánsson and was the first comprehensive catalogue of the Icelandic flora. The earliest attempt to record plants to be found in Iceland was published in 1574[22] but only covered fifty species, especially those of some value to Icelanders, and this interest in medicinal plants was followed in a short list put together in 1638 by Gísli Oddsson, the bishop of Skálholt. The first study by a professional botanist was by Johann Gerhard König, who spent 1764–5 in Iceland and published his work in 1770.[23] This was supplemented by Eggert Ólafsson and Bjarni Pálsson, who included observations on plants throughout their travels. William Hooker provided a list of plants as an appendix to his 1809 journal and various British travellers commented upon the plants they encountered, including Rev. Frederick Metcalfe, who was very particular in noting plant species as he moved from place to place. Compiling plant lists for 'new' places tended to be a feature of nineteenth-century 'exploration'. French interests in the northern latitudes saw a major research initiative in the 1830s and 1840s, with the voyages of the corvette *La Recherche*. In 1835–6 this included Iceland and M. Vahl compiled a list of 432 vascular plants for the country.[24] The extensive list proved a challenge for Mr Charles Babington to check on a short visit in 1846,[25] and in 1860 W. Lauder Lindsay MD, after eight days in the Reykjavík area, produced a new list reviewing all previous records, commenting on the potential limitations of some of Vahl's findings, as it was a 'Liste de Plantes que l'on suppose exister en Islande', and not plants that had actually been seen.[26] Notwithstanding this criticism, Vahl's total of 426 species was not that different from those found by Babington and Lindsay, and not too far from the current number of 480. By the middle of the nineteenth century sufficient time had elapsed to allow non-native plants to become 'settled' in the country such that the overall plant lists were not solely those species that were likely to have been in Iceland before colonization.

A number of plant species were brought to Iceland by the Norse settlers, either intentionally as in the case, for example, of barley, *Hordeum vulgare,* or as weeds, plants that may be termed archaeophytes (the oldest members of the introduced flora, also termed old aliens); there have been attempts to identify these archaeophyte species to reveal the extent of the truly natural flora of the island. Thus species such as caraway, *Carum carvi* (*kúmen*), creeping thistle, *Cirsium arvense* (*þistill*), groundsel, *Senecio vulgaris* (*krossfífill*), and corn spurrey, *Spergula arvensis* (*skurfa*), may be considered examples of the nineteen presently confirmed archaeotypes that were in

Iceland between settlement and 1770, a date based on the records of König, Eggert Ólafsson and Bjarni Pálsson.[27] Several of these plants, as well as native species, had uses for food and medicines, especially as additives to flour to eke out the limited amount that could be produced in most years or that had to be imported. Alpine bistort, *Bistorta vivipara* (*kornsúra*), lyme grass, *Leymus arenarius* (*melgresi*), and scurvy grass, *Cochlearia officinalis* (*skarfakál*), were among those species used as additives, and the last was also used in soups and porridge, with records of it being harvested in late August and September on Grímsey at the Arctic Circle. One of the most prized plants was garden angelica, *Angelica archangelica* (*ætihvönn*), which was used in a wide variety of ways, especially to ease digestive problems and as a restorative. All parts of the plant could be eaten and the roots were dried and traded, being gathered by people going 'mountain rooting'.[28] It was considered sufficiently valuable that the thirteenth-century *Grágás*, or Grey Goose Laws, laid down the possibility of fines or outlawing for anyone caught stealing it. In 1875, having come down off the ice at Vatnajökull, William Lord Watts and his party were delighted to find clumps of angelica, which they devoured hungrily. Mountain avens, *Dryas octopetala* (*holtasóley*), was used for tea and butterwort, *Pinguicula vulgaris* (*lyfjagras*), as a substitute for garlic; common sorrel, *Rumex acetosa* (*túnsúra*), has been a popular and widely used flavouring, good with meat and fish and considered excellent with puffin both in Iceland and the Faroe Islands.[29] The use of some of these plants is now being rediscovered as culinary experimentation looks to the possibilities of including traditional ingredients, and you can buy wine made from crowberry, *Empetrum nigrum* (*krækilyng*), with several different labels available.[30] As part of a return to more natural and healthy food, interest has also turned to the use of seaweed such as dulse, *Palmaria palmata* (*söl*), and carrageen or Irish moss, *Chondrus crispus* (*fjörugras*). Collection rights for dulse were enshrined in law and it was collected and traded, especially with inland farms. Carrageen was prepared into *fjörugrasahlaup*, carrageen moss jelly, for flavouring in sour milk, a far cry from its current popularity as a sashimi garnish.

Apart from plants brought in during the early stages of settlement few other species appear to have successfully arrived until the late nineteenth century and the expansion of trade and tourism. In 1901 Stefán Stefánsson identified 48 species that could be considered alien having arrived between 1840 and 1901, a number that increased to 51 by the crucial third edition of *Flóra Íslands*. In an exhaustive 2013 study, 336 alien species were listed of which 277 were casual aliens (plants that may flourish and possibly breed but eventually die out) and the remaining 59 naturalized aliens, half of all aliens being of European origin.[31] Most of these species are only found in the lowlands and associated with settlements and wasteland or cultivation, although there is some evidence for a small number encroaching into highland areas from certain hotspots, especially around Mývatn and to the west of Vatnajökull at Landmannalaugar and Jökulheimar.[32] The pattern seems to follow the road network, mirroring thousands of years of plant dispersal elsewhere in the world along trade routes associated with people and animals, even transported in dung as animals fed in one area and defecated in another. Apart from the Nootka lupin the only other plant classified as invasive, defined as an alien species that shows a tendency to spread out of control, is cow parsley, *Anthriscus sylvestris* (*skógarkerfill*). This was introduced as an ornamental plant around 1900 and, though seen as less of a problem than the lupin, is expanding – to the detriment of native species; the use of pigs to halt its spread has even been suggested. As yet there does not appear to be a Friends of the Cow Parsley group, although it does colonize areas of Nootka lupin. Despite being ornamental at a garden scale, it does not have the allure of the swathes of blue provided by its fellow alien.

The native Icelandic flora is poor in species with only 480 recorded, of which 0.3 per cent are endemic and the rest characteristic of European floras; anyone familiar with the mountains of the British Isles and Scandinavia would feel very much at home with the plants they would recognize in Iceland, something the renowned British polymath Sir John Herschell had noted in the 1840s. Or as Benjamin Mills Pierce stated in his report in 1860, 'the vegetable wealth of Iceland is not large.' From studies of pollen and macrofossils found in lake sediments and bogs it is clear that there has been little if any change in the flora over the last 10,000 years and that most species were part of the landscape from early in the Holocene. This begs an interesting question and one that has provoked as polarized a debate as the lupin issue – how did Iceland, an island isolated in the middle of the North Atlantic, get its flora; when and from where? Comparison of the species-poor, cold-loving communities of the Holocene with the more warmth-loving and much more diverse flora found in the fossils from the earlier geological history of the country shows a major shift in the ecologies of the plants with western 'American' influences disappearing as the climate became cooler. Although a European source for plants of Holocene age is no longer questioned by most researchers, the means by which these plants arrived and successfully survived in Iceland is still uncertain, with theories falling into two opposed camps. On the one hand is the view that no plants could have survived the last glacial maximum when an ice cap covered the country, leaving a *tabula rasa* or clean slate for plants to colonize; on the other hand is a view that plants could have survived in refugia, ice-free areas from which they could then spread and recolonize as temperatures rose when the ice disappeared.

In pursuit of evidence for the existence of refugia the botanist Steindór Steindórsson identified a number of species found in Iceland that were widely dispersed in Scandinavia, Greenland and North America, but absent from the Alps, and by mapping their distribution across Iceland suggested six possible locations for their glacial survival. At a meeting in Reykjavík in 1962 the refugia hypothesis gained support as representing a plausible explanation for the nature of the Icelandic flora, reflecting contemporary thinking about the comparable situation in Scandinavia, although others disagreed, calling it the 'now merely historical *tabula rasa* theory'.[33] Proponents of the *tabula rasa* theory – which had emerged in the nineteenth century with growing recognition of the scale and severity of past glaciation, and who included the influential geologist Þorvaldur Thoroddsen – argued that a lack of endemic species in Iceland and a growing realization of the impact an enduring Icelandic ice sheet would have had on the possibility for plant survival on nunataks or in hypothesized ice-free enclaves, made the presence of successful refugia highly unlikely. Furthermore if any plants had survived they would not have included many of the species that make up the overall flora, so these would need to have migrated later. As the palaeoentomologist Paul Buckland neatly put it in the title of a paper on the subject, 'If This Is a Refugium Why Are My Feet So Bloody Cold?' Earlier ideas of movement over a land bridge can be easily dismissed given our understanding of recent sea-level history, as the last time a land bridge existed between Europe and Iceland was thought to be around 9 million years ago. Over recent decades advances in palaeoecology, and especially the genetic decoding of plants allowing more precise identification of source areas, has moved the debate strongly in favour of Iceland being a *tabula rasa* with species migrating at the onset and early years of the Holocene. In an exhaustive study of genetic linkages between plant species across the North Atlantic, the northwest European origins of at least 92 per cent of the Icelandic flora are very clearly illustrated, especially for the trees and dwarf shrubs such as birch, juniper and bilberry, putting a further nail in the coffin of extensive species survival in refugia.[34]

Plant migration did, though, still need a mechanism, especially as the European origin of most species meant movement of seeds and plants over a distance of more than 700 km (435 mi.) against the present dominant west–east flow of wind and ocean currents. Charles Darwin had argued for dispersal of plants by icebergs and drift ice from his observations in the southern hemisphere and travellers had been struck by

Logs from northern Russia and Siberia along the shore of northwest Iceland.

the abundance of timber washed up on the shores of Iceland that provided such a valuable local resource. Near Lagarfljót Pike Ward had seen what he termed 'great baulks of timber' around an inland farm. On asking the farmer about their origin he was told that they originated from Siberia and the Arctic Ocean and were deposited at the mouth of the *fljót*. In winter the farmers dug out the wood, especially the larger logs that could reach up to 12 m (40 ft) in length and weighed 2–3 tons, and dragged them on sledges to their farms. Trees floating in the ocean are also a feature of the 1590 Ortelius map identified in a very similar location off eastern Iceland.

With the seasonal reduction of sea ice in the Arctic Ocean and the circulation patterns of currents, it takes several years for timber entering the sea from the northern coasts of Siberia to find its way to Iceland allowing travel from east to west, albeit in an indirect way. The buoyancy of spruce, *Picea*, is only seventeen months, and pine, *Pinus*, and larch, *Larix*, is shorter at ten months; so to get to Iceland trees have to be caught up in sea ice. Rights to driftwood were enshrined in early Icelandic laws and farmers had special marks to use on timber arrived on their land, as it was an extremely valuable resource. Dendrochronological (tree-ring) analysis of current Icelandic driftwood has shown that it all derives from either the White Sea or Yenesei areas of Russia, the youngest taking only six years to arrive in Iceland. Farmers in Iceland had assumed some of the timber was from America, especially that which was termed *rauðviður* or red wood, but this was shown to be larch and of Russian origin.[35] The arrival of timber and plant material on the shores of Iceland is a continuous process, although little if any organic matter remains viable by the time it arrives onshore. James Nicol, quoting Eggert Ólafsson, suggests that northwest Iceland received a range of species from further afield to the west including mahogany, lime, willow, Campeachy wood and cork tree,[36] and that in crashing together the icebergs sometimes produced enough friction to set fire to the timber. Flotsam from the west certainly made it to Iceland, as seen by the occurrence of the heart-shaped smooth nuts or seeds of *Entada gigas* (also called *Mimosa scandens*), monkey ladder. This is a liana, a climbing vine originating from American tropical rainforests, known across the western Atlantic shores by a range of names including sea hearts and Molucca beans. In Iceland they are called *lausnarsteinn* or relief stone as they were used to help women during childbirth, and their tropical origin was identified by Eggert Ólafsson, although they first appear in a written record of 1525.[37]

At the onset of the Holocene with disintegrating ice sheets and ice caps not only over Iceland but also over Scandinavia, and to a limited extent over Scotland, it would perhaps have been possible for movement from Europe to Iceland on a much greater scale than subsequently, as the ocean and atmosphere settled into their new postglacial patterns. Later arrivals would probably have been helped by bird migration as ice-free areas became available in Iceland, Greenland and Arctic Canada, allowing geese in particular to establish seasonal patterns of movement that exist to this day. The genetic study of plants tends to support this concept of matching ocean current and geese migration patterns from the available data, although movement today certainly is less direct than it might have been in the past. On a very much smaller scale birds played an important role in the plant colonization of Surtsey.

Iceland is renowned as either the northern breeding ground for migrating geese or a stopover on the way to Greenland and the Arctic for birds that overwinter in Europe. The importance and scale of breeding areas in the Central Highlands has become apparent over the years, although was likely well known by local farmers for centuries, as geese and swans provided a welcome addition to their diet during the breeding months. The British naturalist and broadcaster Sir Peter Scott, son of the polar explorer Scott of the Antarctic, spent the summer of 1951 counting, catching and ringing pink-footed geese (*Anser brachyrhynchus*) in the area of Þjórsárver south of Hofsjökull, confirming their migration from sites in Britain. His account of the trials and tribulations of what were very early attempts at netting the birds, aided by local farmers acting as guides – published with James Fisher as *A Thousand Geese* – offers a fascinating insight into summer in the deserted Central Highlands in the 1950s where everything had to be done on horseback.[38] Geese can cover the distance

between Scotland and the coast of Iceland in anything from 15 to 22 hours, at an average speed of 60 kph (just under 40 mph), but the real master of rapid migratory travel to and from Iceland has to be the whimbrel which, thanks to satellite tracking, has now been shown to fly non-stop the 5,800 km (3,600 mi.) to its wintering grounds in Africa in as little as five days, travelling at speeds of up to 54 mph (86 kph).[39]

The international importance of Iceland as a migratory centre or stopover for birds is well established. Mývatn and Laxá have been on the tentative list of Icelandic nominations for UNESCO World Heritage Sites, described as a 'unique freshwater ecosystem in the Northern Hemisphere', and the former is well known for the midge fauna from which it gets its name. The midges provide food for the birds and a constant irritation for visitors.[40] Midges are found much more widely in Iceland at many other lakes, including the Mýflúguvatn (Flying-midge lake) that, despite its location at 423 m (1,388 ft) on a plateau in western Iceland surrounded by sparse vegetation, has a local population of voracious flying midges.[41] Mývatn was dredged for its rich deposit of diatomite between 1967 and 2004, a raw material used widely in various forms of filters, but it has retained its biological attractiveness and is believed to have more species of duck than anywhere in the world. In true Icelandic fashion it has a museum with a history. Fuglasafn Sigurgeirs Stefánssonar, Sigurgeir's Bird Museum, is a museum that houses specimens of all except one of Iceland's breeding birds. Overlooking the lake, it is located in a farm that was the home of Sigurgeir Stefánsson, who collected the specimens but drowned in the lake at the age of 37.[42] Friends, family and benefactors raised the necessary money to set up the museum, which was opened in 2008. Archaeological excavations over the last few decades have shown that local communities have successfully sustainably harvested the Mývatn wildfowl population for over a millennium. Written records showed an annual harvest of at least 4,000 eggs a year from a range of duck species as far back as 1712. Numbers were probably under-reported due to avoidance of taxation as by 1941 over 40,000 were taken annually, but recovery of duck shells from three archaeological sites around Mývatn reveals that harvesting of eggs was taking place from the time of the very earliest settlement in the ninth century CE. These excavations also showed that cured fish, as indicated by plentiful partial fish skeletons in the excavated middens, were also a feature of the diet of these early inland settlers, suggesting the existence of a local trade in cured marine products well before the establishment of the Hanseatic trading network.[43]

For those not living in the country there is a tendency to overlook the importance to Iceland of its surrounding seas and what is happening to the north in the Arctic Ocean. We have a view of Iceland as 'North' and in the words of the poet Alyson Hallett need:

> to unhitch the island from a European mind
> connect it with the Arctic's
> southern-most tip[44]

In the worst years of the Little Ice Age, sea ice would drift down from Greenland in the summer and at times encircle the island, reducing temperatures, leading to crop failures and making an already precarious existence all the more prone to collapse. When a period of extended sea ice combined with the eruption of Laki in 1783–4, living conditions were severe enough for the Danish authorities to consider calling for

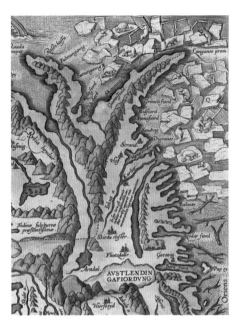

Detail of the Ortelius map, showing both logs in the sea off eastern Iceland and polar bears on the sea ice.

abandonment of the island. The patterns of sea-ice expansion and absence and their implications for climate have been carefully reconstructed for the period since settlement using a variety of sources, principally written records.[45] Again the Ortelius map confirms the existence of sea ice as it is depicted along the eastern coast, but as the ice originates from Greenland it is usually the northwest that is affected first, with ice encircling the island only in extremely bad years. Records for the late sixteenth century, contemporaneous with the production of the map, show increasing sea ice, although the earlier part of the century has poor records and ice appears not to have been nearly as extensive as after 1600 and through to the end of the nineteenth century. The twentieth century and the first two decades of the twenty-first have seen relatively little sea ice around the island.

The day-to-day reality of the effects of sea ice are well illustrated in Pike Ward's account of his time spent in 1905 pursuing his fishing interests.[46] His travels around the country on a variety of steamers were very much dependent on whether the ships could get through drifting ice and access fjords to get to isolated settlements served from the sea rather than over difficult land routes. The impact on the wide range of fishing fleets that exploited the waters around Iceland from Europe and America is also evident, as is the size of the fleets in comparison with the inshore local

Memorial to Captain Floury and his crew of 22 lost off Iceland, 3 April 1877, Chapelle de Perros Hamon, Ploubazlanec, near Paimpol, on the Rue Pierre Loti.

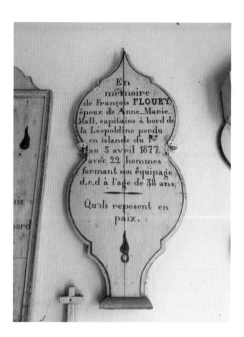

fishing of the Icelanders. Sea ice and the unpredictable and often severe weather experienced in Icelandic waters meant fishing was a dangerous and frequently deadly occupation, as indeed travelling by steamer could be. Rev. Frederick Metcalfe returned to Scotland on the *Arcturus* only days before five steamers were lost in a North Sea gale and *Sölöven*, or Sea Lion, the boat on which Pliny Miles travelled to Iceland, later went down in Faxaflói with only a horse surviving. Pike Ward had barely arrived in Reykjavík before an April storm accounted for 23 fishermen off Viðey within sight of the town. On entering churches in parts of coastal Brittany it is almost commonplace to come across memorials for fishermen lost during the later nineteenth and early twentieth century at the height of the fishing industry, when French schooners, *goélettes*, travelled to Iceland to fish for cod and return with salted fish for the European market. Icelandic cod fishing was central to the economy of a number of northern French towns and villages, such as Paimpol, Dunkerque and Gravelines, immortalized by the author Pierre Loti (Louis Marie-Julien Viaud, 1850–1923) in his 1886 book *Pêcheur d'Islande* (An Iceland Fisherman), which won the Vitet Prize of the Academie Française. At the peak of fishing operations around Iceland there were two to three hundred ships operating from France every summer (Benjamin Pierce reported 269 vessels and 7,000

DESERTS, EROSION, TREES AND CONCERNING SNAKES

sailors operating in 1860) and over the entire period of these fishing operations four hundred were lost, accounting for up to 5,000 deaths. The main cemetery in Reykjavík, Hólavallagarður, has a memorial to French and Faroese fishermen with a quote from Loti.[47] At Fáskrúðsfjörður in eastern Iceland there is a French cemetery with the graves of 49 French and Belgian fisherman, as well as a French hospital and house built during the later nineteenth century at the height of the fishing boom. Five French houses have now been renovated and there is a museum dedicated to the history of fishing by the French and Belgians.[48] There are also road signs in both Icelandic and French, mirroring the roads with Icelandic names in northern France such as the Rue des Islandais Grundarfjördur in Paimpol. The perils of such a way of life, its fragility and the character of its Icelandic context, are best encapsulated in the following from Pike Ward's journal as he travelled in northeast Iceland:

> As we pass Bakki Farm, I notice a cross set up in the tún and find it was the grave of a French sailor, brought onshore when the two vicars were changing and neither was on the spot, so the farmer of Bakki raked in his 20 krónur for allowing the man to be buried on his tún. A simple, rude wooden cross set up, then the poor fellow's name inscribed in rough letters probably done by the captain with black point, marks his resting place until a stray pony or sheep knocks it down and he is forgotten forever.

Sea ice impinging onto the coast of Iceland was indirectly responsible for the last massacre on Icelandic soil, which occurred in October 1615.[49] Large amounts of drifting ice around the northwest of Iceland had led to a very poor summer and three Basque whaling vessels were caught by a September storm and sea ice and driven back into the fjords. Most of the sailors survived and returned to Europe on a kidnapped English boat the following spring but those from the ship of Martín de Villafranca were accused of stealing dried fish at Þingeyri and fourteen were killed and dumped in the sea by the local people, with only one escaping. Following a meeting of the local council under Ari Magnússon from Ögur in Ísafjarðardjúp, the remaining seventeen Basques from this boat, who had planned to overwinter on an island, were declared outlaws and similarly violently, but now legally, killed. Jón lærði Guðmundsson (the Learned), who later wrote his *Natural History of Iceland*, strongly opposed the killings and was forced to flee to Snæfellsnes, providing the basis for a 2008 novel by the 'extraordinary and original' Icelandic writer Sjón (Sigurjón Birgir Sigurðsson) called *From the Mouth of the Whale*, or *Rökkurbýsnir*.[50] It was not until 2015 that the law was formally repealed, leading to the wonderful headline 'Killing of Basques Now Banned in Westfjords', and the massacre is now recognized by a memorial in Hólmavík, which was unveiled at a ceremony including descendants of both Basque and Icelandic families involved.

And so to snakes. The limited flora of Iceland is matched by an even more restricted fauna. There is only one native land mammal, the Arctic fox, *Alopex lagopus*, which is found widely around the Arctic. Polar bears, *Ursus maritimus*, occasionally reach the shores of Iceland on drift ice and are usually quickly shot as they are likely to be starving and considered dangerous. More extensive sea ice in the past meant more regular appearances and although potentially dangerous provided a valuable trading resource in the form of pelts. Another feature of the Ortelius map is fighting polar bears pictured on the drift ice around the northeast of the island. As in Ireland there are no snakes but St Patrick is not invoked as the reason for their absence. James Boswell in his *Life of Johnson* recounts how the doctor boasted of his ability to recite in its entirety Chapter LXXII, *Concerning Snakes*, of Niels Horrebow's *The Natural History of Iceland*. The full chapter is usually taken to comprise 'No snakes of any kind are to be met with throughout the whole island.' Originally the chapter had run to a short paragraph, albeit saying the same, and the relative brevity of the English 1758 version was considered to be due to 'the laconic celebrity to the [unknown] English translator, the author being rather profuse than otherwise'.[51] This chapter is not the shortest in Horrebow's book for Chapter XLII states: 'There are no owls of any kind in the island.' Echoing the terminology of these sources, over a century later Benjamin Pierce in his report said 'there are no indigenous quadrupeds in Iceland.' All other mammals have been introduced, rats perhaps as late as during the eighteenth century and mink in 1931 for fur farming. Escapes soon occurred and it only took 35 years for mink to colonize the whole island.

In 1771 a small herd of reindeer were introduced to Vestmannæyjar by the Danish monarchy, who hoped to improve the quality of Icelandic husbandry.[52] This was followed by larger groups in southwestern Iceland in 1777, northern Iceland in 1784 and Vopnafjörður in eastern Iceland in 1787. Only the eastern group has survived to the present, now comprising a population of up to 7,000 animals. They were never farmed but are hunted to keep a sustainable population and spend most of the year in the highlands away from any human interference, although during the winter occasional reindeer warnings are given when they are forced onto lower ground and become a potential hazard to traffic. Four were killed in a single month in early 2020. The introduction of this alien species was not without its tensions. Following the first appearance of reindeer,

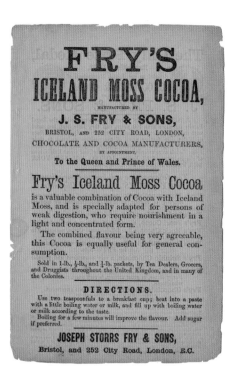

farmers in northern Iceland became concerned that they were responsible
for diminishing stocks of *Cetraria islandica*, commonly known as Iceland
lichen or Iceland moss. This lichen was a valued part of the human diet
and was protected by the mid-thirteenth-century *Grágás*, the Grey Goose
Laws, eaten either as a gruel, with milk or as a tea and believed to have
medicinal properties, as well as providing a way to eke out flour, which
usually had to be bought rather than grown. The flour was used for a
flatbread that reinforced Ida Pfeiffer's general dislike of Icelandic cuisine,
leading the Austrian to undertake several periods of fasting during her
trip. The lichen, known in Icelandic as *fjallagrös* or mountain grass, was
collected in the mountains by expeditions of women and children on
horseback, 'a favourite employment of the females in the summer months'
according to James Nicol in 1840,[33] and stored for use over the winter with
two barrels of moss being considered as valuable as one barrel of flour.
The beneficial properties of Iceland moss were known beyond Iceland and
it was recognized as an 'article of commerce'. In 1859 the English firm of
Dunn and Hewett added it to instant cocoa powder, which Daniel Dunn

had invented in 1820, to make Iceland Moss Cocoa, a drink later taken up by Fry's and Cadbury's. It was marketed by the former as 'Dunn and Hewett's Most Nutritious! Digestible!! And Tonic!!! Beverage'. The source of the Iceland moss in the cocoa is not clear, but small amounts had been exported from Iceland since at least the 1830s and it was still recorded as a commercial good in the later nineteenth century. In recent years it has seen something of a revival, not just as a medicinal agent but also in health foods and as a bread made from *Cetraria islandica* flour; even as *Fjallagrasa*, Icelandic schnapps.[54]

Any species, plant or animal, that threatens the innate character of the Icelandic landscape is likely to prove a focus of concern. From the earliest travellers who encountered a strange and alien landscape to the millions of tourists who currently visit the country, Iceland has been seen as a wilderness – Europe's last wilderness – and anything that threatens the integrity of that image is likely to be treated with caution. The expansion of areas designated as National Parks offers some landscape protection and it is difficult to alter the backbone of the country, its geological heritage, but with increasing economic pressures and a growing global desire to 'experience' Iceland it is likely that the desire to keep the country in as 'natural' a form as possible will be felt more intensely. Coping with alien invasive species, while primarily a technical issue, also unearths deeper social and emotional concerns which in the future will need to be addressed. In debates over the future of the lupin, parallels were drawn in the media with inciting racist attitudes and the danger of engendering a nationalist anti-immigrant identity in the country. If the debate over the Nootka lupin can degenerate into threats and be seen as comparable to racist attitudes against human immigrants, then there is a lot more than science at stake in seeking to conserve the inherent uniqueness of the Icelandic landscape, real or imagined.

CHAPTER 11

ICELAND IN THE ANTHROPOCENE

'THIS MONUMENT IS TO ACKNOWLEDGE THAT WE KNOW WHAT IS HAP-
PENING AND WHAT NEEDS TO BE DONE. ONLY YOU KNOW IF WE DID IT.'[1]

Many earth scientists detest the use of the word Anthropocene and
would not wish to involve themselves in what they see as an academic
navel-gazing exercise that fails to get real geological work done. Two main
questions drive the discussions that have been taking place within geology
regarding the idea of the Anthropocene: have humans had sufficient
impact on the planet to warrant a geological epoch named after them, say
in comparison to the Cambrian explosion or the demise of the dinosaurs?
If the answer to the former is yes, then the second question is: where do
we place the golden spike, the boundary that separates the non-human
from the human, the Holocene from the Anthropocene?[2] As is only to be
expected from any academic debate, both questions are capable of generating considerable argument, at times quite acerbic, and resolution of either
is unlikely to be seen in the near future or perhaps ever. Iceland is a good
place from which to ponder these questions considering its perception as
a relatively untouched environment, that is, predominantly non-human,
where the sheer scale of the natural dwarfs that of the human imprint.
Notwithstanding this contrast, because of its fragility, the sensitivity of
the soils, vegetation and ice and its susceptibility to volcanic eruptions it is
also a location where it should be possible to identify if and when human
influences overcame or at least significantly challenged the natural.

Recent debate on the first question has focused on the minutiae of
what exactly would be required to happen on the planet to generate a separate geological epoch. Opponents of the idea of the Anthropocene argue
that humans are typically overestimating their importance in suggesting
that through their actions enough planetary change has been set in train to
warrant a human epoch. For those against the new derivation the human
impact on the planet would at best qualify for definition as an event in
geological terms, a short-term albeit significant change, but one that in

the fullness of geological time will likely merely merit a footnote in earth history. In its simplest expression, the Anthropocene debate can be summarized as humans either taking a 'narcissistic pride in our importance' or feeling an 'existential despair at our insignificance'.[3] A commentary on this view was provided recently by Adam Frank and Gavin Schmidt, who constructed a thought experiment asking the question: 'Had an advanced civilization existed many millions of years ago would we find direct traces of them in the geological record?' Their conclusion was that because of the paucity of fossils that generally get preserved, the most likely record would be indirect, in the form of anomalous chemical signatures in the rocks.[4]

The so-called Silurian hypothesis, named after an episode of the television series *Doctor Who* rather than the geological period, helps to stoke the concerns of sceptical geologists, but for those prepared to accept the idea of the Anthropocene, Iceland is virtually unique in having had no significant human impact until 874 CE, since when it could be argued, at least in terms of soil erosion, that humans have played a major role in changing the face of the country. Thus in Iceland the magic spike could be precisely dated to when Ingólfur Arnarson first set foot and settled. This post-dates huge impacts from humans elsewhere on the globe, and pre-dates potential markers such as the onset of the Industrial Revolution, the great acceleration of the 1950s and the first release of radioactive isotopes into the atmosphere. It has been suggested that Iceland does indeed have such a golden spike in the geological record in the form of the *Landnám* tephra, an ash layer that was laid down widely across the island immediately before the first settlers arrived, the Landnám or land-take, and is dated to 877±1 CE. In characteristic academic fashion those proposing this have developed a view that such a marker is 'conceptually related to, but distinct from, the global Anthropocene'[5] – researchers on Iceland seeking distinctiveness for the country even in the midst of a major global debate. It is not the aim here to delve any deeper into this potential geological quagmire but to take advantage of the informal idea of the Anthropocene as a way of signposting what may happen in the future to the landscape of Iceland, and to make such an assessment in the context of likely future human (anthropogenic) influences and actions. Thus, while not an apologia for the Anthropocene, the lure of the terminology and the intellectual context it offers could not be avoided, offering a perspective that is not simply scientific but requires a broader cultural, social and economic framing.

Projections of climate trends for the twenty-first century have been made by the Icelandic Meteorological Office, Veðurstofa Íslands, using a range of models and pathways for greenhouse gas emissions.[6] These give a

ICELAND IN THE ANTHROPOCENE

leave the known safety of home to cross possible *jökulhlaup* pathways to get to designated safe locations. Some living in the shadow of Katla feel that so long as they are above the levels affected by flooding they feel safer staying put – 'there is not a better place to be' – and would fear a major snow storm and avalanches more than an eruption.[10] Given the loss of life in the twentieth century due to avalanches compared to eruptions this is a very logical and rational approach.

For most of Iceland, due to decreasing ice volumes and as a long-term response to the earlier disappearance of the large IIS, isostatic recovery will continue and the land will rise at up to 3 cm/1¼ in. a⁻¹ in some parts, assisting the erosive capacity of the rivers and slowly enhancing the impressive nature of the many canyons and incised river valleys. Estimates of uplift between the years 2000 and 2100 around southern Vatnajökull vary between 2.5 and 5 m (8¼–16½ ft) depending on the rate of ice thinning on the ice cap.[11] The harbour of Höfn, the langoustine capital of Iceland, is threatened by this rise, which is leading to a shallowing of Hornafjörður, potentially preventing access by fishing vessels. In the southwest around Álftanes and Seltjarnarnes the plate boundary is actually sinking, with a cessation of volcanic activity requiring replacement of rock sea walls to protect the settlements close to the shore.

On Sunday 18 August 2019, Iceland held a ceremony to recognize the demise of the Ok glacier. Hundreds of people walked to the former site of the glacier to see the installation of a plaque marking its passing, including scientists, the geologist Oddur Sigurðsson, who had been overseeing monitoring of the extent of Icelandic glaciers for several decades, the UN High Commissioner for Human Rights Mary Robinson and President Katrin Jakobsdóttir.[12] In 1890 the glacier covered 16 km² (6¼ mi.²) but by 2012 it had shrunk to 0.7 km² (¼ mi.²), and by 2014 it was officially declared to be no longer active. The plaque, in Icelandic and English, is brief but to the point:

A letter to the future
Ok is the first Icelandic glacier to lose its status as a glacier.
In the next 200 years all our glaciers are expected to follow the same path.
This monument is to acknowledge that we know
What is happening and what needs to be done.
Only you know if we did it.
Águst 2019
415ppm CO_2

Satellite image of Ok glacier lying to the west of Langjökull, 14 September 1986.

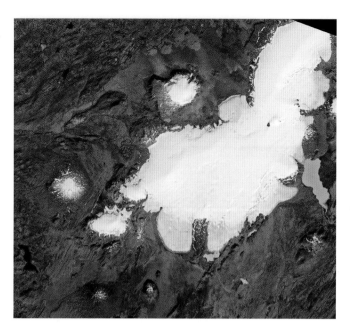

Satellite image of same area, 1 August 2019, showing the disappearance of the former glacier.

Snæfellsjökull, possibly the next glacier in Iceland to disappear.

OPPOSITE
Jökulsárlón proglacial lake in front of Breiðamerkurjökull that will continue to expand with further warming.

This was not just a local event as it had global coverage and a month later several hundred people held a similar glacier 'funeral' for the Pizol glacier in the Swiss Alps.[13] Between 2000 and 2020 Iceland lost 800 km² (308 mi.²) of its glaciers and according to climate projections for the remainder of the twenty-first century, based on the highest levels of continuing greenhouse gas emissions it is believed that Iceland could see the further loss of 25–35 per cent of its ice cover and that the 30-metre-thick (98 ft) ice that covers Snæfellsjökull in the west will effectively disappear by mid-century. The smaller sensitive ice caps of Hofsjökull and Langjökull could be reduced by 60 per cent and 85 per cent by 2100, when Vatnajökull could be the only ice cap to retain a summer covering of snow. Under the RCP of 8.5 (Recommended Concentration Pathway – a measure of projected greenhouse gas emissions) Drangajökull in the northwest would disappear by 2050, or certainly by 2100 even under the lower emissions scenarios (lower RCPs).[14] Should retreat of the Breiðamerkurjökull outlet of Vatnajökull continue as predicted, which since the 1930s has seen the appearance and growth of the famous glacier lake Jökulsárlón, it will reveal a buried fjord extending 300 m (985 ft) below sea level, further allowing warmer ocean water to enter the lake, accelerating calving of the ice front, creating a new landscape and probably providing an even more spectacular sight for visitors. Within the last few years part of Breiðamerkurjökull has developed a 1.1-kilometre-long (¾ mi.) split, threatening to help break up part of the ablation area. Loss of ice on a longer timescale will lead to a

Fjallsárlón proglacial lake in front of Fjallsjökull at the western end of Breiðamerkurjökull.

broader change in the surface of the country with a landscape dominated by subaerial lava and ash deposition rather than sub-glacial tuffs and palagonites, reminiscent of the country in the Tertiary prior to the expansion of the current ice caps.

There is an inevitable tendency to accept a somewhat catastrophic view of an ice-free Anthropocene in Iceland, a land named after all from its glacial heritage. Presidents have expressed their concern over the plight of glaciers in recent years. Vigdís Finnbogadóttir, in her introduction to Helgi Björnsson's book *The Glaciers of Iceland: A Historical, Cultural and Scientific Overview*, argues that 'Glaciers are an essential part of the Icelandic identity – all true Icelanders are fascinated by glaciers from childhood.' Halldór Laxness, in the novel *World Light* (*Heimsljós*), wrote about 'Where the glacier meets the sky, the land ceases to be earthly; no sorrows live there anymore . . . beauty alone reigns there.'[15] Nevertheless, in her study of the communities in Hornafjörður M. Jackson found a more ambivalent and nuanced attitude to disappearing ice. Families, especially farmers that had lived under the constant threat of advancing ice and glacier floods, welcomed the increased security to their homes and livelihoods brought by retreating glaciers with new land appearing, land that had previously been farmed and exploited by the first settlers that was overrun by Little Ice Age advances – 'a long twilight where some Icelanders may savor some beneficial ripples of glacier loss'. Expansion of the local tourist industry into a year-round experience through exploiting visits to ice caves being discovered along the retreating ice fronts has served to reduce migration of the younger part of the population and brought money into the local economy. In 2015–16, 60,000 visitors visited Breiðamerkurjökull over the winter, when access to the caves is considered safest.[16] Although not at the top of the global list, diminishing Icelandic glaciers may be marketed as part of the developing trend for 'last chance tourism' – places and sights to see before they disappear.[17]

The loss of ice will initially prove something of a bonus for hydro-electric power generation as the increased run-off provides an enhanced continuous energy source. By 2030 there will be a 30 per cent increase in run-off but as the glaciers diminish in size so this too will reduce. One other result of shrinking glaciers is the exposure of unsafe bounding cliffs, which, as pressure is released due to the removal of ice, become increasingly unstable. The 2018 warning regarding potential failures at Svínafellsjökull might be expected to become a much more common occurrence. Earth movements, or *berghlaup*, can be seen all around Iceland. Many probably occurred early in the Holocene as the ice disappeared and unstable slopes adjusted to new conditions. Once aware of how to identify these features it is amazing just how many can be seen across the landscape, especially in the west and east on the higher steep slopes of valleys cut through by ice during the last glacial period. In Iceland slope mass movements, landslides and avalanches have been responsible for more deaths over the last century than any other natural phenomenon. In the twentieth century, 193 people died, 69 of these after 1974, and 34 in two separate incidents in 1995 in northwest Iceland at Súðavík and Flateyri.[18] This is very much a continuous threat. In January 2020 fishing boats in the harbour at Flateyri were destroyed by an avalanche that crossed a protective barrier and two months later houses had to be evacuated due to a further avalanche threat. The steep, narrow fjords of northwest Iceland are very susceptible to such a hazard as settlement is restricted to strips of flatter land on the lowermost areas below the steep slopes and extreme snow events can occur, building up unstable snow depths. Protection features can now be seen around the main settlements considered at risk, as for instance above Ísafjörður. With future uncertainties over climate trends it is difficult to see how this threat will develop. In a sparsely populated area such as Iceland many avalanches and rockslides will not impact on people. July 2018 saw one of the biggest historical rockslides in Hítardalur in western Iceland, but this was only heard by a local fox hunter and later discovered where it had covered pasture and temporarily blocked a good salmon-fishing river with a deposit 20–30 m (65½–98½ ft) thick.[19] Restoring the fishing is estimated to cost over 1SKr1m, around $730,000. Increased contrasts between cold and warm weather episodes allowing for high snowfalls and rapid thaw will increase the likelihood of potentially dangerous avalanches and landslides, as will the spread of settlement onto less safe areas of land. The sensitivity of Icelandic mountain permafrost to changing surface temperatures means that active layers are likely to deepen, and with continued seismic triggering this will also add to the threat.[20] Iceland is unlikely to become a safer place in the higher and steeper mountains.

The image of a country blessed by natural power resources yet still recognized as pristine and a wilderness, a rare occurrence as we move into the Anthropocene, might imply that Iceland is unlikely to change in any noticeable ways over the next century, apart from its loss of ice. There are, however, increasing pressures on the landscape directly arising out of its geological, visual and experiential qualities. With demand increasing for renewable power there is pressure to expand the areas used for hydropower, which contributes over 70 per cent of Icelandic energy supplies. Gullfoss may have been saved but the Kárahnjúkar power plant and its reservoirs, completed in 2009, transformed 1,000 km² (385 mi.²) of wilderness in the eastern Central Highlands and as demand continues to grow with the possibility of a submarine energy supply to Europe, other areas considered suitable for power generation will be threatened. The account by former model Heiða Guðný Ásgeirsdóttir, co-written with the internationally well-known Icelandic poet and novelist Steinunn Sigurðardóttir, of her return to sheep farming and the campaign she had to wage to prevent a power company from taking over the valley in which her farm was located, *Heiða: A Shepherd at the Edge of the World*, reflects this conflict at the local scale.[21] The 2018 film *Woman at War* (*Kona fer í stríð*) similarly focuses on the actions of a single woman, Halla, seeking to prevent the destructive effects of the aluminium industry, the user of much of the new power generation on the Icelandic highlands.[22] The latter has typically thoughtful Icelandic idiosyncrasies in the film – the recurring sight of a trio of Icelandic musicians, a drummer, sousaphone player and accordionist playing alone in various of the locations is one – but at its heart lies a very real issue of how even renewable, clean energy has its problems if we really value 'nature', even more so if the concept of wilderness lies at the heart of one's view of the intrinsic value of the Icelandic landscape.

Tension between power generation and environmental protection is not just a very recent phenomenon in Iceland. Attempts to expand energy capacity on the Laxá river near Mývatn in the late 1960s and 1970s led to the blowing up of a dam at Miðkvísl in August 1970.[23] Over a hundred local farmers claimed to have been involved although no one was charged and this produced a major rethink of government policy, saving the local wetlands. There is archive footage of the activities in the area that has recently been made into a film, *Hvellur*, the English title of which is *The Laxá Farmers*, rather than the more literal translation of *Bang*.[24] It also prompted an essay from Halldór Laxness warning of threats to the environment, 'The War against the Land' ('Hernaðurinn gegn landinu').[25] The roots of the Laxá conflict lay in a more complex debate than simply development against conservation. The original damming of the Laxá in 1938–9 had been undertaken by Laxárvirkjun, a company owned by the town of Akureyri, and brought electricity to the town, not to the local communities. Building bigger dams and generating more energy, which would have serious local environmental and economic effects, was seen as an extension of distant power, not just from Akureyri but also from central government and Reykjavík, paying little regard for local democracy. The parallel between this period and current concerns is marked, only this time the power and the pressure comes from outside the country, from trading blocks and corporations, and has a global dimension. Recognition of the importance of engagement between nature and humans has been a feature of recent political developments in the country, with the foundation of a successful Left-Green party, Vinstrihreyfingin-grænt framboð, described in English as an eco-socialist party whose chairperson, Katrin Jakobsdóttir, became Prime Minister in 2017. Given the complex system of political parties in Iceland this does not necessarily mean that the country follows a strong all-encompassing green agenda, but environmentalism would be expected to have an important voice in government policy.

Calculations of carbon budgets for the globe now firmly show that volcanic production of CO_2 is dwarfed by human emissions. Only between 28 and 60 million tons are annually produced volcanically as against the 10 gigatons that can be ascribed to humans using fossil fuel resources. Four per cent of this natural production comes from Katla according to measurements undertaken in 2016–17, a figure that was higher than previously estimated but still minuscule on a planetary scale. Even a cat-astrophic geological event such as the Chicxulub impact 66 million years ago barely produced half as much CO_2 as we do at present. Compared to other countries Iceland does not therefore have a problem with its carbon budget but is looking to use its geological resources to lessen its impact even further. The CarbFix project seeks to mineralize CO_2 and store it in basalt.[26] At the geothermal power plant at Hellisheiði they remove the CO_2 from the steam that has been used to generate power in turbines, add water and then return it below ground where it is mineralized and stored in the basalt. One of the geologists, Edda Aradóttir, has been quoted as saying 'we have enough basalt to deal with all fossil fuel available. Theoretically it can solve the problem.' CarbFix2 is now in progress, supported by the EU, and was visited by Angela Merkel in August 2019, so is considered a technology of high potential for carbon capture and storage on an international scale. The most recent results from the project have shown 90 per cent of CO_2 to be mineralized within two years of injection.

The clash between environmental protection and economic exploitation does not only revolve around power generation. When Hákon Bjarnason first introduced the Nootka lupin he argued that increasing the diversity of the flora of the country could only be of benefit in making for a more productive land and that there was a clear economic imperative for using beneficial alien species; he made the same case for the Siberian larch. Prevention of erosion of the precious land surface was paramount, but at what point should reclamation stop or be held in check? An early slogan used by those arguing for using the lupin was '*Græðum landið*' ('Healing the land'), and there is no doubt that there was a strong moral imperative to halting the expansion of eroded areas. Once the balance is tipped between healing areas seen to be in need of repair and reclaiming land that would

Landmannalaugar, one of the most environmentally sensitive yet popular tourist areas in Iceland.

normally take years to develop a soil cover, such as glacier forelands and *sandur* plains, or might never gain a vegetation cover on a Holocene times-cale, as in some areas of the deserts of the Central Highlands, the need for management has to be considered. Once active management is undertaken then the mystique of the natural is lost, a growing characteristic of the Anthropocene. Had the lupin not had such photogenic blue flowers, is it likely that it would have had so many admirers? And can it now be thought of as enhancing the natural experience of wilderness areas?

A central concern for the future has to be the exponential growth in tourist numbers and the pressure this puts on the landscape, especially the focus on perceived honeypot locations. Film location tourism has put pressure on a number of sites, some extremely fragile and sensitive, to add to the overpopulation of key areas such as Þingvellir, Geysir and Gullfoss. As a result of a steamy sex scene in Season 3 of *Game of Thrones* the popular swimming cave at Grjótagjá near Mývatn, a former possible hideout for the sixteenth-century outlaw Jón Markússon, had to be closed in July 2018 due to the lack of respect shown to the site by tourists according to the landowners. The number of film tourists is likely to increase in the future as in a post-Coronavirus world Iceland is seen as a safe location within which to film, especially as the larger film companies already have a good knowledge of what the landscape can offer. The number of people diving and snorkelling in Silfra at Þingvellir has increased from 20,000 in 2014 to over 70,000 in just four years, and after a fifth death in seven years diving was briefly suspended in the spring of 2017.[27] What has been termed a 'tourism epidemic' has led to concern over the lack of a clear strategy for

MOST UNIMAGINABLY STRANGE

how to deal with accommodating increasing numbers and the conservation needs of a sub-arctic landscape where, for example, off-road driving can lead to irreparable damage to the thin soil and vegetation cover.[28]

In any dystopian view of the Anthropocene the idea of preserving wilderness becomes of major importance yet defining what wilderness is has long been a matter of contention across different countries, partially bound up with different views of what should constitute national parks. For Iceland, wilderness areas were originally uninhabited, uninhabitable wastelands, the home of outlaws and supernatural beings. With the move towards independence and under the influence of Romanticists they became examples of the sublime and the essence of the country – wilderness as 'more a social *idea* than a *reality*'.[29] Bjarni Thorarensen, Iceland's pioneer Romanticist poet, while studying in Denmark in the early nineteenth century considered the Danish landscape to be like a face without a nose or eyes, whereas in Iceland the face was whole, the glaciers and mountains providing the eyes and nose – a complete landscape. In the twentieth and the twenty-first centuries pressures on what are considered wilderness areas are not just local and national but global. To some extent, people, not just in Iceland but globally, are trying to come to terms with ideas of just how wild wilderness has to be and how best to manage it, as ignoring these issues may not result in their survival. Approaches that fail to recognize the cultural importance of wilderness areas, defined within the terms of the people who actually live in or adjacent to them, are unlikely to flourish. Iceland has more than one word for wilderness: there is *öræfi* or wasteland, desert, and, more akin to the English term, *víðerni* or *ósnortið víðerni*, unspoiled wilderness, as well as *óbyggðir*, land that is unsullied by humans – used in the name of the Wilderness Center, or *Óbyggðasetur Íslands*, set up in 2015 on the edge of the Central Highlands as a base for exploring the wilderness of the centre of the country. Icelandic law also formally defines wilderness in a 2013 Act as, among other things, an area of land

> of not less than 25 km² in size, or such that one can enjoy solitude and the natural landscape without disturbance from human structures or traffic resulting from mechanized vehicles . . . that is situated at a distance of at least 5 km from human structures and other technical traces.[30]

The importance of solitude, emptiness and quiet are themes that have recurred through travellers' accounts. Among them James Bryce in 1872 talked of a 'strange stern beauty, stilling the soul with the stillness of nature';[31] Elizabeth Jane Oswald at a similar time considered that the landscape 'gives awful calm to the spirit';[32] Bunsen wrote to his mother of 'the desolation and death-like quiet that prevails in this Icelandic mountain country'; and it is peace, quietness and stillness that often characterize what we would see as intrinsic to wilderness areas. Enshrining such qualities in legislation is challenging; gone are the days when Ebenezer Henderson could comment, 'One may travel 200 miles without perceiving the smallest symptom of animated being of any description whatsoever,' and Bunsen was able to observe that 'the only living things we met on this whole trip were a sea gull that had lost its way and two ptarmigan.' Managing wilderness areas is challenging but managing the idea of wilderness is even more difficult.

While images and the written word will always be important in delivering ideas of landscape it is noticeable in Iceland that the Kárahnjúkar project saw musicians take an important role in protest at the loss of wilderness and threats to the landscape. Björk set up the *Náttúra* environmental campaign as an attempt to protect nature and support local and sustainable products. Sigur Rós played at the protest camp on their *Heima* tour and their music is considered by some to be deeply rooted in the Icelandic landscape, just like Ólafur Elíasson's installations. The joint *Náttúra* concert held by Björk and Sigur Rós in Reykjavík in 2008 attracted 30,000 people, 10 per cent of the population of Iceland. The *Heima* tour celebrated the diversity of the country and their environmental concerns are particularly apparent in the video *Vaka*, released in 2003, which is set in a post-apocalyptic world with children in layers of clothes and gas masks playing in a schoolyard covered in black ash falling from a red sky.[33] The film *Draumalandið* (Dreamland), produced in 2009, has a soundtrack by Valgeir Sigurðsson that specifically focuses on the above issues with the Kárahnjúkar project as its target, one reviewer seeing it as 'a visual treatise that exposes the greed and corruption that has befallen Iceland . . . it speaks for ecology through stunning cinematography and the pacing of a practiced orchestra.' The film is based on the book by Andri Snær Magnason, *Draumalandið. Sjálfshjálparbók handa hræddri þjóð* (2006), published in English in 2008 as *Dreamland: A Self-help Manual for a Frightened Nation*.[34] It is to be hoped that the Anthropocene will not come to a dystopian end but issues of planetary protection, what the music academic Nikki Dibben has called a sense of planet rather than a sense of place, will remain at the heart of political debates and wilderness areas, however defined, will come under more and more pressure, either to be exploited for necessary economic gain, or to be protected for necessary spiritual gain, aside from their inherent geological and biological importance.

The Icelandic landscape of the Anthropocene will not be that different from the landscape that has inspired and enthralled visitors, explorers, tourists and scientists over the centuries, and which developed the close affinity and respect between Icelanders themselves and their immediate surroundings. The growing physical demands of tourism and energy are unlikely to abate, leading to concerns about just how to retain that almost mythic otherness of the country which underlies much of its magnetic draw for those not fortunate enough to be born there. The perceived safe experience of travelling across the dystopian volcanic and glacial landscapes that can be found almost immediately on leaving any settlement lies at the heart of the experiences tourists have of any visit, and the clean, almost virtuous character of the range of renewable energy resources that the juxtaposition of the geology and latitude provide are similarly central to the twenty-first-century identity of the country. Confronting these potentially conflicting aspects lies at the heart of Iceland in the Anthropocene, whatever geological label you decide to use.

POSTSCRIPT

The persistence of Iceland both physically and emotionally is something
that all who have an interest in the country experience. Curiosity and an
appreciation of the sense of other that the country elicits recur not only
throughout the writings of travellers, explorers and artists but in the way
scientists have come to understand the landscape. The *otherness* of Iceland
has challenged visitors in different ways. Ebenezer Henderson was able to
take refuge in the Bible as he sought to comprehend what he encountered.
William Morris, Sabine Baring-Gould and W. G. Collingwood could
rely on the sagas to offer them an understanding of how people could
best come to terms with life in this hard land. Scientists such as Robert
Bunsen, Sir Joseph Banks and Hans Ahlmann, and informed interested
visitors such as Uno von Troil, Sir George Mackenzie, Henry Holland and
many others, could revert to what they understood was the true route to
knowledge through careful observation, meticulous measurement and the
generation of theories for others to consider and test. Artists and writers
sought in their own way to find the essence of the land. For Icelandic
scientists, writers and musicians, what to foreigners is 'otherness' is the
everyday; glaciers, volcanoes, hot springs, boiling mud and idiosyncratic
culinary tastes were and are facts of life, the nature of the country, nothing
exotic. All, though, seem to recognize that an encounter with the Icelandic
landscape cannot be simply a matter of seeing and understanding in a
straightforward, objective, scientific way – a more satisfying awareness
comes with broadening the mind and accepting a need to see the country
through a variety of different critical prisms.

    The prisms through which we have been able vicariously to view the
Icelandic landscape provide us with what the geologist Marcia Bjornerud
describes as 'a clear-eyed view of one place in Time, both the past that
came before us, and the future that will elapse without us'. She goes on
to argue how understanding the morphology of a particular landscape
is similar to the 'rush of insight one has on learning the etymology of an
ordinary word. A window is opened illuminating a distant yet recognizable

past.' Iceland is no 'ordinary word' but by seeing the landscape from a wide spectrum of viewpoints, hopefully it has been possible to gain a better insight into a unique and fascinating country.[1] One feature that has consistently emerged over the centuries is a sense of awe that has accompanied any engagement with the landscape, whether in those seeking to enjoy the experience for its own sake or those looking for a deeper geological understanding. As the great popularizer of science Carl Sagan said in 1996, 'scepticism and wonder are skills that need honing and practice.'[2] There is no better place to practise them than in Iceland, where 'the landscape is a story, full of hostility and strange turns, but habitable too.'[3]

Returning from a visit to Iceland leaves one with a sense of loss and envy of people who are native to the land and who understand it with a greater depth and affinity than those of us who flit in and out and merely scratch the surface of what the landscape offers and requires. In days when getting to Iceland proved an adventure in itself, there was often a sense of regret at leaving, realizing that there may not be another opportunity to visit the country. For an earth scientist it is not just the physical expression of processes and forms that is the draw but the deeper emotions that the landscape generates that matter. In the preface to his book on Skaftafell the geomorphologist Jack Ives reflects such a view in justifying writing about his work and the Skaftafell area as a whole: 'If only a few experience a fraction of the enthusiasm, joy and self-fulfilment that I have found there, then the effort will have been amply worthwhile.' From my own experiences in Iceland I can only echo his sentiments and hope that this book goes some way towards reflecting my own joy at the privilege of being able to study such a fascinating and intriguing landscape.

REFERENCES

The references are designed to provide sources for the material used in the book and occasionally include notes that complement the text in a brief discursive form. An attempt has been made to find references in English wherever possible, either as translation or in the original, although very occasionally material may only be available in Icelandic. Where multiple quotations have been used from single authors such as Ida Pfeiffer, Lord Dufferin or William Morris, they have not all been fully referenced. The reader is referred to the original reference unless they originate from a different work by the author. Page numbers of the quotations have not been provided, as in many cases they may vary from edition to edition.

INTRODUCTION

1 Unless stated otherwise all the quotes from William Morris in this book are taken from *The Collected Works of William Morris, with Introductions by His Daughter May Morris*, vol. III: *Journals of Travel in Iceland, 1871–1873* (London, 1896). Like many of the earlier references quoted in the book it is now available in facsimile form, in this case based on the 1911 edition from Elibron Classics published in 2005.

2 Ófeigur Sigurðsson, *Öræfi: The Wasteland* (Dallas, TX, 2018).

3 Halldór Laxness, *Brekkukotsannáll* (Reykjavík, 1957), first translated into English as *The Fish Can Sing* in 1966, but now English quotes are usually from the Magnus Magnusson translation of 2001.

4 Various conventions have been followed in the book regarding forms of dating. In historical contexts CE and BCE (Common Era and Before Common Era) are used in preference to AD and BC, but because of the large number of dates covering the last millennium and up to the present (indeed also into the future), the use of CE for any dates after 999 CE has been dropped. Thus 1000 (AD or CE) becomes simply 1000.

5 This was an important seventeenth-century book used widely and having the self-explanatory title *Geography Anatomiz'd; or, A Compleat Geographical Grammar: Being a Short and Exact Analysis of the Whole Body of Modern Geography after a New and Curious Method Whereby Any Person May in a Short Time Attain to the Knowledge of That Most Noble and Useful Science*. It was first published in 1699/1700 and ran to many editions, succeeding in its titular aim as far as many travellers were concerned.

6 A number of sources provide analyses of the extant accounts from this period. In particular Karen Oslund, *Iceland Imagined: Nature, Culture, and Storytelling in the North Atlantic* (Seattle, WA, 2011), and

Sumarliði R. Ísleifsson and Daniel Chartier, eds, *Iceland and Images of the North* (Quebec, 2011). In the latter the paper by Sumarliði R. Ísleifsson, 'Islands on the Edge: Medieval and Early Modern National Images of Iceland and Greenland', pp. 41–66, provides an extremely valuable summary.

7 Passages from Ida Pfeiffer are taken from the 1852 English translation: Madame Ida Pfeiffer, *A Visit to Iceland and the Scandinavian North* (London), which is available as an e-book of the 1853 second edition (www.gutenberg.org).

8 Perhaps the most illuminating anecdote is to be found in T. Ross Browne, *The Land of Thor* (New York, 1870). Towards the end of his trip across Russia, Scandinavia and Iceland he enjoyed local hospitality at a farm near Reykjavík. When it was time to retire he was followed to his room by the only daughter, who showed no signs of leaving him until he was thoroughly prepared for bed. After some time watching as he began to remove his clothes, she eventually took 'a position directly in front . . . seized hold of the pendent casimere and dragged away with a hearty good-will. I was quickly reduced to my natural state with the exception of a pair of drawers . . . in a very ragged condition . . . I whirled into bed, and the young woman covered me up and wished me a good night's sleep . . . so ended this strange

and rather awkward adventure.' This is a story also found in accounts by other visitors and begs the question of whether, reading about such an experience, later visitors were prepared for this eventuality and expected such hospitality, making sure it appeared in their final account. Rev. Frederick Metcalfe, for instance, talked of the daughter of the farm pulling off his 'inexpressibles', and Lord Dufferin's doctor companion Fitz experienced the same near Geysir when he left the group due to a cold and stayed alone on a farm.

9 John Barrow Jr, *A Visit to Iceland, by Way of Tronyen, in the 'Flower of Yarrow' Yacht, in the Summer of 1834* (London, 1835). Lieutenant John Barrow (1808–1898), who eventually became the head of the Admiralty Record Office, was the son of Sir John Barrow, one of the founders of the Royal Geographical Society in London, a passionate supporter of Arctic exploration.

10 The problems associated with Blefken's account have become a focus for academics examining the earlier accounts of Iceland, and were recently resurrected when on 1 July 2015 *Die Welt* printed an excoriating extract about Iceland written by the German author Oliver Maria Schmitt under the title 'Iceland Is as Large and as Dead as East Germany'; among other observations he called the

island 'obnoxious' and the people 'crazy and delusional'. Reactions to this in Iceland, and especially to the author ('A German Comedian? That's a Contradiction in Terms') were not complimentary and tellingly he was described as 'no Blefken', reflecting a country with a real awareness of its written past. (Magnús Sveinn Helgason, 'Who Is This German That Has Offended So Many Icelanders and International Icelandophiles?', *Iceland Magazine*, 2 July 2015.)

11 The journals of Sir Joseph Banks that included his account of travelling in Iceland were not published as such, although researchers have had access to the original material over the years. In 2016 the Hakluyt Society eventually published *Sir Joseph Banks, Iceland and the North Atlantic: Journals, Letters and Documents*, edited by the Icelandic historian Anna Agnársdóttir, the most complete collection of the writings of one of the best known British Icelandophiles.

12 William Jackson Hooker (1785–1865), later Sir William Hooker and Director of the Royal Botanic Gardens at Kew, did publish an account of his travels: William Hooker, *Journal of a Tour in Iceland, in the Summer of 1809* (London, 1813), which is available online as part of the Wellcome Collection (www.wellcomecollection.org). The self-proclaimed Protector of Iceland Jørgen Jørgensen, known

as *Jörundur hundadagakonungur* (the Dog-day King), wrote an account of his life which shows his facility for self-aggrandizement in the title: *The Convict King, being the Life and Adventures of Jorgen Jorgensen, Monarch of Iceland, Naval Captain, Revolutionist, British Diplomatic Agent, Author, Dramatist, Preacher, Political Prisoner, Gambler, Hospital Dispenser, Continental Traveller, Explorer, Editor, Expatriated Exile and Colonial Constable retold by James Francis Hogan* (London, 1891).

13 All quotes from Henderson are based on the second edition of 1819. The original was published as: Ebenezer Henderson, *Iceland; or, The Journal of a Residence in That Land, during the Years 1814 and 1815. Containing Observations of the Natural Phenomena, History, Literature, and Antiquities of the Island; and the Religion, Character, Manners and Customs of Its Inhabitants* (Edinburgh and London, 1818) and is available online at www.hathitrust.org. The American version produced in 1831 was an abridged copy; one of the reasons for this was given as: 'there are many passages in it, that can be of little importance to any body.'

14 Quoted by Paula Bracken Wiens, 'Fire and Ice: Clashing Visions of Iceland in the Travel Narratives of Morris and Burton', *Journal of the William Morris Society*, 11 (1996), pp. 12–18.

15 George E. J. Powell of Nanteos in

Wales was a man of independent means who travelled and collected extensively during his life. On his death he bequeathed all his collections to Aberystwyth University: 'In my will, therefore, I had left to your University – as well as being quite the worthiest and most intelligent corporate body in my dear but benighted town – all I possessed "of bigotry and virtue"'.

16 W. Gershom (1854–1932) was an artist and Professor of Fine Arts at University College, Reading, and the son of a watercolourist. He was a friend of John Ruskin and Arthur Ransome and lived in the English Lake District. His Icelandic journey was originally published as *A Pilgrimage to the Saga-steads of Iceland by W. G. Collingwood and Jón Stefánsson* (Ulverston, 1899), and more recently in 2013 published by the R. G. Collingwood Society through Cardiff University, as *W. G. Collingwood's Letters from Iceland: Travels in 1897 by W. G. Collingwood and Jón Stefánsson*. A W. G. Collingwood archive at Cardiff University includes both his papers and artworks.

17 The term 'thanatourism' derives from thanatopsis, usually interpreted as a consideration of death, which has its origins in a poem written in its final version in 1821 by the American William Cullen Bryant (1794–1878). Although dark tourism now encompasses a range of experiences from natural disasters to genocides, it is argued that places such as Iceland were attractive for similar dark reasons in the nineteenth century: K. Walchester, 'The British Traveller and Dark Tourism in Eighteenth- and Nineteenth-century Scandinavia and the Nordic Regions', in *The Palgrave Handbook of Dark Tourism Studies*, ed. P. R. Stone et al. (London, 2009), pp. 103–24.

18 The Fiske Collection can be accessed online at Cornell University Library, www.library.cornell.edu, and there are also useful commentaries and blogs on various aspects of the collection.

19 Emily Lethbridge, 'Americans in Sagaland: Iceland Travel Books, 1854–1914', in *From Iceland to the Americas*, ed. Tim William Machan and Jón Karl Helgason (Manchester, 2020), pp. 137–59.

20 Karl Grossman, 'Across Iceland', *Geographical Journal*, 3 (1894), pp. 261–81.

21 Review in the *Times Literary Supplement* for 7 August 1937. A more sympathetic review might have been expected from Sidney Barrington Gates, as he was an aeronautical scientist as well as a man of letters who had poems published by Virginia and Leonard Woolf's Hogarth Press.

22 Quoted in Gary L. Aho, 'William Morris and Iceland', *Kairos*, 1 (1982), pp. 102–33.

23 A Google search of this topic produces literally pages of hits from different sites, all of course covering the same spread of

locations. One of my strangest experiences in taking student groups to Iceland was to visit the area in front of Svínafellsjökull in the early 2000s and puzzle over the origins of patches of white plastic lying among the moraines. It was fake snow left from *Batman Begins*, released in 2005.

24 Lyrics can be found at https://boniver.org.

25 See www.artangel.org for an overview of the Vatnasafn installation.

26 A full overview of Ólafur Elíasson's work is on his website at www.olafureliasson.net.

27 Michael Tucker, 'Not the Land, but an Idea of Land', in *Landscapes from a High Latitude: Icelandic Art, 1909–1989*, ed. Julian Freeman (London, 1989). This was the first British exhibition of twentieth-century Icelandic art and it came about through a collaboration between the National Gallery of Iceland (Listasafn Íslands) and Brighton Polytechnic Gallery. Apart from an Introduction by Magnus Magnusson, then at the height of his TV career, there are a range of chapters and some excellent reproductions of paintings. The title echoes Lord Dufferin's *Letters from High Latitudes*, published in 1857 and which Magnus Magnusson argued opened the eyes of the British to Iceland, or Ultima Thule. The excellent exhibition in Brighton did not have quite the same effect as this over a century later.

28 *The Life and Times of Henry Lord Brougham Written by Himself in Three Volumes* (New York, 1871). Lord Brougham went on to become Lord Chancellor, invented the Brougham four-wheeled horse-drawn carriage and was a keen supporter of improvements in education; he was also an enthusiastic patron of Cannes.

29 One of the most popular travel accounts, it was first published by John Murray in 1857 as *Letters from High Latitudes: Being Some Account of a Voyage in 1856 in the Schooner 'Foam' to Iceland, Jan Mayen and Spitzbergen* (London, 1857). In 1934 it was published by Oxford University Press, as part of their World's Classics series, having previously been published by J. M. Dent and Sons in their Everyman's Library series in 1910. All quotations used here are from the 1934 edition. Lord Dufferin was an example of the pre-steamship English visitor, usually from the gentry, who arrived by private yacht and managed to leave an impression that anyone from England coming to Iceland was likely to be rich, a problem for less well-off later nineteenth-century British travellers in Iceland.

30 *The Iceland Journal of Henry Holland, 1810* was eventually published by the Hakluyt Society in 1987, edited by Andrew Wawn. It was the first published English account, although the manuscript journal was known of by some

travellers. It had been translated into Icelandic by Steindór Steindórsson and published in 1960 as *Dagbók í Íslandsferð eftir Henry Holland* (Reykjavík).

31 Sir George Steuart Mackenzie Bt, *Travels in the Island of Iceland, during the Summer of the Year* MDCCCX (Edinburgh, 1811). Online access is available through the Hathi Trust at www.hathitrust.org.

32 *Travels in Iceland by Eggert Ólafsson and Bjarni Pálsson Performed 1752–1757 by Order of His Danish Majesty*, first published in English in 1805, but originally published in 1772 as *Reise igiennem island. Ferðabók Eggerts Ólafssonar og Bjarna Pálssonar um ferðir þeirra á Íslandi árin 1752–1757*, and since available in a variety of facsimile forms. Eggert Ólafsson (1726–1758) also wrote poetry and is considered an early Icelandic nationalist. Bjarni Pálsson (1719–1779) went on to become Iceland's Surgeon General in 1762.

33 Although there was no official journal published by Sir Joseph Banks there was a series of contributions from party members published in 1780 as *Letters on Iceland: Containing Observations on the Civil, Literary, Ecclesiastical, and Natural History; Antiquities, Volcanoes, Basaltes, Hot Springs, Customs, Dress, Manners of the Inhabitants, &c., &c., Made during a Voyage Undertaken in the Year 1772, by Joseph Banks, Assisted by Dr. Solander, Dr. J. Lind, Dr. Uno von Troil, and Several Other Literary and Ingenious Gentlemen*. The Letters of Uno von Troil were published separately in different forms such as that produced by Cambridge University Press in 2011.

34 Quoted in Anna Dóra Sæþórsdóttir, C. Michael Hall and Jarkko Saarinen, 'Making Wilderness, Tourism and the History of the Wilderness Idea in Iceland', *Polar Geography*, 34 (2011), pp. 249–73.

35 Douglas Hill Scott, *Sportsman's and Tourist's Handbook to Iceland* (Leith, 1899).

36 Published as *Mitteilungen der Vereinigungen der Islandfreunde, Organ der Vereinigung der Islandfreunde* with a frontispiece featuring a rather stern image (of an eagle?) that was very much reflective of its time. The Association of Friends of Iceland was founded in Dresden on 15 March 1913 and disbanded in 1936–7 due to attempts by the Nazi regime to use it for its own political ends, and published its journal half-yearly free for members. A revised Association was set up after the war in 1950 in Hamburg (www.islandfreundehamburg.de).

37 Reference to this debate is occasionally to be found in geology textbooks and academic papers tend to be very detailed; in this case Wikipedia provides reasonable coverage of both schools of thought.

38 Gustav Georg Winkler (1820–1896) visited Iceland in 1858 and wrote two books on the country, of which *Island. Der Bau seiner*

Gebirge und dessen geologische Bedeutung (Munich, 1863) was the most obviously Neptunist in character.

39 The contribution of Sveinn Pálsson is thoroughly discussed in Ólafur Ingólfsson, '"On Glaciers in General and Particular" . . . The Life and Works of an Icelandic Pioneer in Glacial Research', *Boreas*, 20 (1991), pp. 79–84.

40 More extensive comments on Larsen and other painters of the time can be found in Karen Oslund, *Iceland Imagined: Nature, Culture and Storytelling in the North Atlantic* (Seattle, WA, 2011).

41 By 2019 there had been a reduction in overall numbers but they still exceeded 2 million, details of which can be found on the website of the Iceland Tourist Board, Ferðamálastofa, www.ferdamalastofa.is.

42 Rev. Frederick Metcalfe, *The Oxonian in Iceland; or, Notes of Travel in That Island in the Summer of 1860 with Glances at Icelandic Folk-lores and Sagas* (London, 1861).

43 Rev. W. T. McCormick, *A Ride across Iceland in the Summer of 1891* (London, 1892).

44 William Lord Watts, *Across the Vatna Jökull; or, Scenes in Iceland: Being a Description of Hitherto Unknown Regions* (London, 1876). It is available as a reproduced facsimile in several forms. The *Arcturus* was built at Dumbarton on the Clyde as the *Victor Emmanuel* and origi-nally destined for the Mediterranean. It was bought by Koch and Henderson and renamed to sail as a mailship between Denmark and Iceland, a route subsidized by the Danish government, making its first crossing on 17 April 1858. It sank on 5 April 1887, following a collision with a British steamship, but all on board were rescued.

45 William George Lock, *Guide to Iceland: A Useful Handbook for Travellers and Sportsmen* (London, 1882). He appears to have self-published from 16 Kingston Terrace, Charlton, southeast London.

46 Charles S. Forbes, *Iceland: Its Volcanoes, Geysers and Glaciers* (London, 1860). Forbes (1829–1876) was a naval Commander who travelled to Iceland on a whim, meeting a friend by chance in London who was setting off for Leith and so decided to join him. When asking his friend what to bring, he said, 'Your oldest clothes, water-proof suit, gun and fishing-rod'.

47 The *Sportsman's and Tourist's Handbook to Iceland* appears to have been produced annually for the steamship passengers and Scott acknowledges William Lock for allowing him to use parts of his earlier Guide. He did recommend visiting Gullfoss but in the 1900 edition access to the waterfalls was not considered possible as the ferry was no longer operating, although he does not say which ferry this refers to.

While acknowledging the general excellence of Lock's information he does point out that with regard to fitting out ladies for a visit to Iceland 'Mr. Lock was somewhat at sea with this.'

48 Mrs Alec Tweedie (Ethel Brilliana Harley), *A Girl's Ride in Iceland* (London, 1889), second edition including Preface, 1895. She was one of the first British women to travel in Iceland, following in the footsteps of Ida Pfeiffer. She did have an unnamed woman friend with her for the visit.

49 Richard F. Burton, *Ultima Thule; or, A Summer in Iceland*, 2 vols (London, 1875).

50 Andrew Wawn, 'William Morris and Translations of Iceland', in *Translating Life: Studies in Translational Aesthetics*, ed. Shirley Chew and Alistair Stead (Liverpool, 1999), pp. 253–76.

51 Janet Powney and Jeremy Mitchell, 'The Lure of Iceland: The Northern Pilgrimages of Mary Gosden (Mrs. Disney Leith)', *The Victorian*, 6/1 (2018). Another remarkable woman traveller, Mary Gosden (1840–1926) married General Robert William Disney Leith, 32 years her senior, by whom she had six children. On his death in 1892 she taught herself Icelandic and with various of her children made fourteen trips to Iceland, wrote a number of books, especially for children (for example in the series *Peeps at Many Lands*), and translated *Njáls saga* (*Brennu-Njáls saga*).

52 Alice Selby, *Icelandic Journal*, ed. A. R. Taylor (London, 1974). This is the publication by the Viking Society for Northern Research of Alice Selby's 1931 pencil-written journal of her journey undertaken while a lecturer at Nottingham University.

53 Sabine Baring-Gould, *Iceland: Its Scenes and Sagas* (London, 1863). This is now available in an excellent format from Signal Books (Oxford, 2007), as part of its Classic Travel Writing series.

54 Pike Ward, *The Icelandic Adventures of Pike Ward*, ed. K. J. Findlay (Exeter, 2018).

55 Extracts of reviews were appended to the second edition of Mrs Tweedie's book, which were very positive, if at times a little condescending – calling it a 'charming little book' (*Athenaeum*), 'compact' (*Graphic*) and 'unaffected and charming' (*Truth*), and its author 'a plucky young lady' (*Pictorial World*); several included comments such as 'the authoress learnt to ride like a man' (*Spectator*) and 'she rode her pony, in a masculine attitude . . . 160 miles, in four days, which, for a lady, seems to us to be a remarkable feat' (*Illustrated London News*). Mrs Tweedie went on to edit an autobiography of her surgeon father and several novels, including *Life in a Highland Shooting Box*, described by one reviewer as 'racy and readable'. Just over a

decade after Mrs Tweedie's riding exploits, Mrs Disney Leith rode 480 km (300 mi.) across Iceland in 1901 at the age of sixty.

56 See Sæþórsdóttir, Hall and Saarinen, 'Making Wilderness, Tourism and the History of the Wilderness Idea in Iceland'.

57 Andrea de Pollitzer-Pollenghi, 'The Vatnajökull', *Alpine Journal*, 48 (1936), pp. 257–79. The 1930s saw an expansion of motor travel in Iceland and the pre-Second World War mix of horses and motors is well illustrated in Olive Murray Chapman's *Across Iceland*, published in 1930. In her account she describes in detail a horseback ride around the Snæfellsnes peninsula, undertaken against local advice, and subsequent sections of her journey on overcrowded vehicles along roads that she would have much preferred to travel in the saddle.

58 Hans Wilhelmsson Ahlmann, *Land of Ice and Fire* (London, 1938).

59 Jack D. Ives, *Skaftafell in Iceland: A Thousand Years of Change* (Reykjavík, 2007).

60 The map by Abraham Ortelius is very well known in cartographic circles and was probably produced in 1585, before being printed for the first time in 1590 as *Additamentum IV* to his *Theatrum Orbis Terrarum*. Over the next two decades copies were produced with information on the reverse in a variety of languages. Original copies were in black and white

but most that survive have been hand-coloured, possibly not that long after publication. A more detailed discussion of the map can be found at https://islandskort. is, produced by Landsbókasafn Íslands – Háskólabókasafn.

61 Niels Horrebow (1712–1760) was a Danish lawyer and judge who visited Iceland in 1749 when falling upon somewhat straitened circumstances. His account was originally published in Danish in 1752 as *Tilforladelige efterretninger om Island*, translated into English in 1758 as *The Natural History of Iceland: Containing a Particular and Accurate Account of the Different Soils, Burning Mountains, Minerals, Vegetables, Metals, Stones, Beasts, Birds, and Fishes*. It is available in facsimile copy, one reprint weirdly having a cover featuring a disused railway line, apparently because it is part of a series.

62 *Uppdráttr Íslands* – Mapping Iceland – was the outcome of surveys undertaken by Björn Gunnlaugsson (1788–1876), details of which can be found at https:// islandskort.is, appearing first in 1844 and then published in 1850 in a smaller form. Björn Gunnlaugsson was born in rural northern Iceland and played an important role in Icelandic mathematical education, as well as using his time outside in tents while surveying to write an extensive philosophical poem – *Njóla*, or Night. He was termed 'The Wise Man with a

Child's Heart', with 'a reputation of being a very strange man indeed', in a review of his work by Benedikt S. Benedikz (*Scandinavian Studies*, 75 (2003), pp. 567–90).

63 Edward Weller (1791–1884) began producing maps for publication in *Cassell's Weekly Dispatch* in serialized form in 1855, allowing readers to build up an atlas in instalments. At its height 60,000 copies were sold every week.

64 Details of the project ICECHANGE can be found at www.svs.is.

65 *Sjálfstætt fólk* was published in two parts, the first in 1934 and the second in 1935. The first English translation as *Independent People* was produced in 1945.

66 Jón Kalman Stefánsson's trilogy of novels was published in Icelandic between 2007 and 2011, and in English by the Maclehose Press between 2010 and 2015 (preceded by editions in French and German). His 2013 book *Fiskarnir Hafa Enga Fætur* was translated as *Fish Have No Feet* in 2016 and nominated for the Man Booker International Prize.

67 Auden had been attracted to Iceland by his father's interest in the country, as well as his belief that he was of Icelandic descent, and after attending lectures at Cambridge by J.R.R. Tolkien. The book was the subject of a contract agreed with Faber & Faber for a travel book, destination unspecified. For MacNeice it was part escapism but also arose out of a fascination for islands already demonstrated in his literary work.

68 Simon Armitage and Glyn Maxwell, *Moon Country: Further Reports from Iceland* (London, 1996). Their journey in the footsteps of Auden and MacNeice took place in 1994. Like *Letters from Iceland* this was published by Faber & Faber.

69 The quote is from the American Pliny Miles's account of his travels, presumably in 1853, as the book is written and published in the summer of 1854 – *Norðurfari: Rambles in Iceland* (New York).

70 Jón Steingrímsson, *Fires of the Earth: The Laki Eruption, 1783–1784* (Reykjavík, 1998). This is the first English translation of his account.

71 Browne, *The Land of Thor*.

72 Andrew James Symington (1825–1898) was a Glasgow merchant, poet, biographer and antiquary. He was on the same sailing of the *Arcturus* as Charles Forbes in 1859. His son, also called Andrew James, left Scotland in 1882 and set up one of the most famous port-producing estates in the Douro valley in Portugal.

73 Geir Zoëga (1830–1917) was indeed a very highly prized guide who was used by Pliny Miles as well as John Ross Browne. When requested by the well-known American traveller Bayard Taylor in 1874 he was otherwise engaged guiding the king of Denmark's party but arranged for his nephew, another Geir, to assist him. In 1889 he was very much the principal guide in

Reykjavík; Mrs Alec Tweedie had to go through him to organize guides for a trip to Geysir and into the twentieth century he was still named in handbooks such as the *Sportsman's and Tourist's Handbook to Iceland* by Douglas Hill Scott, in which he had an advertisement showing him also to be a merchant. He offered guides speaking English, Danish and French, as did Thorgrímur Gudmundsen, who had been guiding since 1873 and had testimonials from H. Rider Haggard and the prince of Hesse, among others.

74 Tucker, 'Not the Land, but an Idea of Land'.

75 Tempest Anderson preferred photographing active volcanoes and collected an impressive quantity of images now curated in the Tempest Anderson Photographic Archive at the Yorkshire Museum in York. A short biography can be found on the website of the Yorkshire Philosophical Society (www.ypsyork.org), his favourite organization for his slide shows.

76 James Todd Uhlmann, 'Gas-light Journeys: Bayard Taylor and the Cultural Work of the American Travel Lecturer in the Nineteenth Century', *American Nineteenth Century History*, 13 (2012), pp. 371–401. The inclusion of Ida Pfeiffer in his *Cyclopaedia of Modern Travel* was interesting considering Bayard Taylor was considered by Uhlmann as something of a 'model of liberal/republican masculinity'.

See also Tom F. Wright, 'The Results of Locomotion: Bayard Taylor and the Travel Lecture in the Mid-nineteenth-century United States', *Studies in Travel Writing*, 14 (2010), pp. 111–34.

77 Comment in the *New York Times* obituary for Pliny Miles, 4 May 1865.

78 Part of the Fiske Collection. Some photographs by Frederick W. W. Howell were included in his book, *Icelandic Pictures Drawn with Pen and Pencil* (London, 1893), published for some reason by the Religious Tract Society. They were advertised as 'The Best and Most Comprehensive Collection in Existence' in the *Sportsman's and Tourist's Handbook to Iceland*, and could be obtained from the photographer at Handsworth in Birmingham.

79 Images of Willem van de Poll's photographs taken in Iceland in 1934 are available online at several sites, with an especially good collection at www.lemurinn.is.

80 Images of paintings by both Ásgrímur Jónsson and Jóhannes Sveinsson Kjarval can be found online. Ásgrímur Jónsson has a collection at his house, which was opened in 1960, two years after his death. It is now part of Listasafn Íslands, the National Gallery of Iceland. Jóhannes Sveinsson Kjarval was born in the Borgarfjörður region and there is now a Kjarval Experience that describes him as Iceland's favourite artist (www.icelandtravel.is).

81 Tucker, 'Not the Land, but an Idea of Land'.

82 Quoted by Paul Laity in an article in *The Guardian* in April 2011 entitled 'Eric Ravilious: Ups and Downs', www.guardian.com.

83 Tucker, 'Not the Land, but an Idea of Land'.

84 Benedikt Gröndal (1826–1907), 'a latter-day Icelandic Renaissance man', made this comment in the 1870s.

85 The report was mainly concerned with trying to support the case for buying Greenland, out of a concern over the expansion of British interests through the new Dominion of Canada. It was also argued that Denmark had neglected Iceland and that the Icelanders were looking to the United States for support – something echoed by Pliny Miles, although he was also looking to Americans to settle in Iceland and bring with them 'productive and practical arts'. By comparison Metcalfe talked of what he saw as the 'parental affection' of the Britons for Iceland given their common heritage.

86 Ahlmann, *Land of Ice and Fire*.

87 Philip Marsden, *The Summer Isles* (London, 2019).

88 Charles Darwin, *On the Origin of Species by Means of Natural Selection; or, The Preservation of Favoured Races in the Struggle for Life* (London, 1859).

CHAPTER 1

1 Madame Ida Pfeiffer, *A Visit to Iceland and the Scandinavian North* (London, 1852).

2 Charles S. Forbes, *Iceland: Its Volcanoes, Geysers and Glaciers* (London, 1860). Forbes (1829–1876).

3 Lord Dufferin, *Letters from High Latitudes. Being Some Account of a Voyage in 1856 in the Schooner 'Foam' to Iceland, Jan Mayen and Spitzbergen* (London, 1857). There is a recent biography of Lord Dufferin which concentrates on his political career: Andrew Gailey, *The Lost Imperialist: Lord Dufferin, Memory and Mythmaking in an Age of Celebrity* (London, 2015); appropriately produced by John Murray, who published the first edition of Dufferin's book in 1857.

4 Holocene is used to describe the period after the last Ice Age or cold stage, and although its age is often rounded off to the last 10,000 or 10,500 years, using ages in radiocarbon years, the more correct timing is 11,700 calendrical not radiocarbon years. This is based on the boundary as defined by the International Commission on Stratigraphy in 2018, specifically 11,700 calendar yr b2k (11,700 calendar years before 2000 CE), from the North Greenland Ice Core Project (NGRIP) ice core in Greenland. In some texts the more general term Postglacial is used. Thus the correct description for the Holocene is the last 11,700 years of earth history.

Confusions can arise because as a unit of time a radiocarbon year does not directly relate to a calendrical year, but is based on amounts of radiocarbon produced that have varied over time.

5 Pfeiffer, *A Visit to Iceland and the Scandinavian North*.

6 There are several books that cover the basic geology of Iceland. The original classic is Þorleifur Einarsson, *The Geology of Iceland* (Reykjavík, 1994), first published in 1991 as *Myndun og mótun lands – jarðfræði*. Thor Thordarson and Armann Hoskuldsson, *Iceland* (Harpenden, 2002), is part of the Classic Geology in Europe series, and probably the best as a guide around the island. Focused on the tourist route is Agust Gudmundsson, *The Glorious Geology of Iceland's Golden Circle*, published in 2017. There are two briefer popular guides: Ari Trausti Guðmundsson and Halldór Kjartansson, *Earth in Action: An Outline of the Geology of Iceland* (Reykjavík, 1996), and its companion volume Ari Trausti Guðmundsson, *Volcanoes in Iceland: 10,000 Years of Volcanic History* (Reykjavík, 1996). Apart from being a geologist, Ari Trausti Guðmundsson stood for President of Iceland in 2012 and is a member of the Alþingi (Iceland's national parliament), representing the Left-Green Movement, having at one stage in his career been Chairman of the Icelandic Communist-Unity (Marxist-Leninist) party. He was one of the team responsible for developing the ice cave at Langjökull. Following in the tradition of earlier Icelanders introduced in this book he has several other interests and skills, including writing fiction and being a well-known mountaineer. He is, like many Icelandic earth scientists, multi-lingual. The Icelandic publication *Jökull: The Icelandic Journal of Earth Sciences*, formerly the journal of the Iceland Glaciological Society, publishes on Icelandic geology, and a 2008 volume on *The Dynamic Geology of Iceland* is an excellent source of reasonably up-to-date research. The latest contribution on Iceland's geology is Tamie J. Jovanelly, *Iceland: Tectonics, Volcanics and Glacial Features* (Hoboken, NJ, 2020), but this retails at around £150. For the purposes of this book a range of these sources have been used and specific references are only used where there are particular pieces of research that warrant recognition.

7 Details at www.katlageopark.com.

8 There is a defined area of volcano-speleology that hosts an annual symposium on the subject, and a text on the range of features that occur in association with lava tubes is Charles V. Larsen, *An Illustrated Glossary of Lava Tube Features* (Vancouver, 1993).

9 A good summary that gives a flavour of Clarence Dutton is

Antony R. Orme, 'Clarence Edward Dutton (1841–1912): Soldier, Polymath and Aesthete', *Geological Society, London, Special Publications*, 287 (2007), pp. 271–86.

10 Christina Sunley, *The Tricking of Freya* (New York, 2009).

11 Kirsten Hastrup, 'Icelandic Topography and the Sense of Identity', in *Nordic Landscapes: Region and Belonging on the Northern Edge of Europe*, ed. Michael Jones and Kenneth R. Olwig (Minneapolis, MN, 2008), pp. 53–76. The full quote is: 'By comparison to Versailles or Persepolis, the scenery is austere and architecturally very modest; yet the whole place is tremendously impressive and symbolically loaded with Icelandness.'

12 There is a biography of Jónas Hallgrímsson in English: Dick Ringler, *Bard of Iceland: Jónas Hallgrímsson, Poet and Scientist* (Madison, WI, 2002).

13 Milan Kundera, *Ignorance: A Novel* (London, 2002), first published in French in 2000 as *L'ignorance*.

14 Halldór Laxness, *The Atom Station* (London, 1961), first published in Iceland as *Atómstöðin* on 21 March 1948, completely selling out on the day of publication.

15 Charles Lyell, *Principles of Geology: Being an Attempt to Explain the Former Changes of the Earth's Surface, by Reference to Causes Now in Operation*, was originally published by John Murray in three volumes in 1830, 1832 and 1833.

16 There are a number of dating techniques used in geology which rely on the radioactive decay of isotopes of various elements. K-Ar is the decay of ^{40}K, an isotope of potassium, to stable argon. Because the half-life of ^{40}K is so long, it can be used to date suitable rocks such as lavas over very long timescales, well in excess of 100,000 years.

17 Palaeomagnetic dating does not provide a specific age for a rock or sediment directly. Because the location of the North Pole is constantly changing, the palaeomagnetic signal relative to the location of the Pole found in rocks and sediments provides a pattern over time. These patterns can be matched by comparing sequences from different series of rocks or sediments and, supported by other dating methods such as K-Ar, they can provide estimated ages for sequences and hence individual elements of that sequence.

18 These recognized abbreviations for dates are used thus: Ma means a million years, and so 2 Ma is 2,000,000 years; ka is used for thousands, so 2 ka means 2,000 years.

19 A brief overview of Walker's impressive contribution to geology can be found in David Pyle, 'The Legacy of George Walker', *Mineralogical Magazine*, 73 (2009), p. 1051. There is also now ICEPMAG, the Iceland Palaeomagnetism Database,

which currently holds almost
10,000 lava palaeomagnetic data
points. A more personal flavour
can be seen from his obituaries.
The Guardian included the
comment, 'Those who had the
good fortune to be with him in
the field carry fond memories of
a man in his element, sharing his
acute observations and insights
with a twinkle in his eye to an
audience struggling to keep up
with him physically and mentally.'
See also Scott R. Rowlands and
R.J.S. Sparks, 'A Pictorial Summary
of the Life and Work of George
Patrick Leonard Walker', in *Studies
in Volcanology: The Legacy of George
Walker*, ed. T. Thordarson et al.,
pp. 371–400, which includes field
sketches and photographs of his
fieldwork in Iceland.

20 Bunsen's ideas were indeed well
ahead of his time as far as rhyolites
and the implications for possible
mixing of magmas was concerned.
See Ray E. Wilcox, 'The Idea
of Magma Mixing: History of a
Struggle for Acceptance', *Journal of
Geology*, 107 (1999), pp. 421–32.

21 As for instance in James Nicol, *An
Historical and Descriptive Account
of Iceland, Greenland, and the Faroe
Islands, with Illustrations of Their
Natural History* (Edinburgh, 1841).

22 The term Tertiary is no longer
generally accepted but is to be
found across a range of sources.
For Iceland there are two epochs
within the period covered by the
Tertiary that are relevant: the

Miocene (23–5.3 Ma) and Pliocene
(5.3–2.6 Ma). The Tertiary covered
the period from 66–2.6 Ma. Now
the most accepted terminology is
Palaeogene for 66–23 Ma followed
by the Neogene to the present,
although see Reference 25.

23 Details of this eruption and all
other major Icelandic eruptions
can be found on the Institute
of Earth Sciences website at the
University of Iceland (Hásko-
li Íslands), which also has a
constantly updated Catalogue of
Icelandic Volcanoes. The university is also home to NordVulk, the
Nordic Volcanological Center. See
www.nordvulk.hi.is.

24 Overviews of this can be found in
the introductory geology texts for
the country; more specific details
are in academic journals, many of
which unfortunately may not be
on open access. A good summary is in Friðgeir Grímsson and
Leifur A. Símonarson, 'Upper
Tertiary Non-marine Environ-
ments and Climatic Changes
in Iceland', *Jökull*, 58 (2008),
pp. 303–14. There is a very
detailed academic coverage of
this topic in Thomas Denk et al.,
*Late Cainozoic Floras of Iceland: 15
Million Years of Vegetation and
Climate History in the Northern
North Atlantic* (Dordrecht, 2011).

25 There has been a lot of geological
debate about the acceptability
of the Quaternary as a separate
geological unit of stratigraphy
but it is now recognized as a

Period with the onset at 2.6 Ma (actually 2.588 Ma at the base of the Gelasian Stage, the accepted GSSP (Global Stratotype and Section Point), found in Italy). The Quaternary includes the Pleistocene and the Holocene, but for all pre-Holocene events the two terms of Pleistocene and Quaternary are effectively interchangeable.

26 A detailed account is: Jón Eiríksson, 'Glaciation Events in the Pliocene-Pleistocene Volcanic Succession of Iceland', *Jökull*, 58 (2008), pp. 315–30.

27 The remains of mines are now flagged for tourists, as for example at Sýðridalur in western Iceland; see www.westfjords.is.

28 See Reference 24; it is possible to download Chapter Two of Denk et al., which covers the history of research in this area, through www.academia.edu.

29 Leo Kristjánsson, 'Iceland Spar and Its Legacy in Science', *History of Geo- and Space Sciences*, 3 (2012), pp. 117–26.

30 Details of the museum are at www. steinapetra.is.

31 See Ingi Þ. Bjarnason, 'An Iceland Hotspot Saga', *Jökull*, 58 (2008), pp. 3–16. There is also a very extensive and well-sourced Wiki-

pedia entry for 'Iceland Hotspot'.

32 Andrew Wawn, *The Vikings and Victorians: Inventing the Old North in Nineteenth-century Britain* (Cambridge, 2002).

33 Co-author with W. G. Collingwood of *A Pilgrimage to the Saga-steads of Iceland*, he spent much of his life in London and during the Second World War lobbied Churchill, arguing that Iceland would prefer British occupation.

34 Yet another dating convention! BP is used to define ages Before Present. This terminology derives from its use in radiocarbon dating where all ages were defined BP with Present being 1950 (CE). With increasing age, 'Present' will of course rather strangely recede but as a general term it has sufficient precision and allows direct comparison of different ages of sediments or organisms.

35 Alyson Hallett and Chris Caseldine, *Six Days in Iceland* (Stoke Gabriel, 2011).

36 See Introduction, Reference 67.

37 See www.visithafnarfjordur.is.

38 For a curious exposure to the topic it is worth trying Bego Antón, *The Earth Is Only a Little Dust Under Our Feet* (2018).

39 See www.archello.com/project/harpa.

CHAPTER 2

1 Inscription on the 1590 Ortelius map over the erupting Mount Hekla. The translation is the one that is generally to be found associated with

any representation of the original Latin.

2 Neither Jakob Ziegler nor Caspar Peucer visited Iceland, but both were involved in developing the

evolving theology of Protestant-ism, alongside their geographical interests. See Jelle Zeilinga de Boer and Donald Theodore Sanders, *Volcanoes in Human History: The Far-reaching Effects of Major Eruptions* (Princeton, NJ, 2002), and Richard Handler, ed., *Central Sites, Peripheral Visions: Cultural and Institutional Crossings in the History of Anthropology* (Madison, WI, 2006).

3 Arngrímur Jónsson hinn lærði, the Learned (1568–1648) was much aggrieved by what he saw as 'the intolerable errors of foreigners'; hence his publication of *Brevis commentarius de Islandia* and his counterblasts to Blefken and Peerse (see Introduction), but he failed to stop many of the myths being reproduced in later geographies. His book *Crymogæa* (literally iceland, from the Greek), which was published in Hamburg in Latin in 1609, proved more successful and introduced European scholars to the extensive literary legacy to be found in Iceland.

4 The *New York Times* obituary for Pliny Miles, while generally favourable, tended to damn his writing with faint praise. His account of his visit to Iceland was 'a pleasant book of the gossipy sort', written by 'a plain but forcible writer, depending upon a laborious array of facts, rather than rhetorical effort. In person he was a striking figure . . . wearing a beard that never scraped acquaintance with a razor.'

5 Charles S. Forbes, *Iceland: Its Volcanoes, Geysers and Glaciers* (London, 1860).

6 James Nicol, *An Historical and Descriptive Account of Iceland, Greenland and the Faroe Islands with Illustrations of Their Natural History* (Oxford, 1840). James Nicol (1810–1879) was a Scottish geologist who, after studying Arts and Divinity at Edinburgh, changed his interests to geology while studying in Germany before eventually becoming Professor of Geology at Aberdeen. His 1840 book was almost certainly a compendium of previous accounts with no real evidence that he visited Iceland.

7 John van Wyhe, *Wanderlust: The Amazing Ida Pfeiffer, the First Female Tourist* (Sydney, 2019), a long overdue biography of a truly remarkable woman traveller.

8 Helga Schutte Watt, 'Ida Pfeiffer: A Nineteenth-century Woman Travel Writer', *German Quarterly*, 64 (1991), pp. 339–52.

9 The most quoted figure is thirty but occasionally higher numbers have been used, for example 32 at www.icelandvolcanoes.is, the online Catalogue of Icelandic Volcanoes. A clear description and map of the systems and their relationship to volcanic zones can be found in Thorvaldur Thordarson and Ármann Höskuldsson, 'Post-glacial Volcanism in Iceland', *Jökull*, 58 (2008), pp. 197–228, and also in

T. Thordarson and G. Larsen, 'Volcanism in Iceland in Historical Time: Volcano Types, Eruption Styles and Eruptive History', *Journal of Geodynamics*, 43 (2007), pp. 118–52.

10 The definitions of zones are also not consistent across sources, with the online Catalogue of Icelandic Volcanoes only defining four. Of the seven zones, four are simply termed 'zones', whereas the Mid-Iceland Belt connects the zones while the other two, the Öræfajökull and Snæfellsnes Belts, are intraplate belts. These structural differences are quite clear on any map of the volcanic zones.

11 See Thorvaldur Thordarson and Ármann Höskuldsson, 'Postglacial Volcanism in Iceland', T. Thordarson and G. Larsen, 'Volcanism in Iceland in Historical Time: Volcano Types, Eruption Styles and Eruptive History', and Guðrún Larsen and Jón Eiríksson, 'Holocene Tephra Archives and Tephrochronology in Iceland: A Brief Overview', *Jökull*, 58 (2008), pp. 229–50.

12 Paula Bracken Wiens, 'Fire and Ice: Clashing Visions of Iceland in the Travel Narratives of Morris and Burton', *Journal of the William Morris Society*, 11 (1996), pp. 12–18.

13 The displays at the Surrey Zoological Gardens continued on a near annual basis until 1851 (when they were displaced by the Great Exhibition); the mural of Iceland and Mount Hecla [*sic*] was followed in 1841 by one of Rome. The conservatory was 90 m (300 ft) in circumference and had 560 m^2 (6,000 ft^2) of glass, housing a number of wild animals during its heyday. Brief details of the displays can be found at https://vauxhallhistory.org and also in *The Literary World: A Journal of Popular Information and Entertainment* of Saturday 25 May 1839 (Price 2d!).

14 Plinian eruptions have been some of the most powerful, capable of ejecting ash over 11 km (6.8 mi.) into the upper atmosphere. They are named after the 79 CE eruption that destroyed Pompeii and Herculaneum, which killed the scholar Pliny the Elder and was described by his nephew Pliny the Younger. The most recent major Plinian eruption globally was the 1991 eruption of Mount Pinatubo in the Philippines, which impacted global climate over the next one to two years.

15 Noted in Haraldur Sigurdson, 'Sigurdur Thorarinsson, 1911 [*sic*]–1983', *Eos*, 64 (1983), p. 450.

16 There are two excellent obituaries/appreciations: Helgi Björnsson, 'Obituary Sigurdur Thorarinsson, 1912–1983', *Journal of Glaciology*, 29 (1983), pp. 521–3, and Richard S. Williams Jr, 'Sigurður Þórarinsson, 1912–1983: An Appreciation', *Polar Geography and Geology*, 8 (1984), pp. 85–8. There is also an interview with him in 1979 available on YouTube that is extremely

illuminating about his life and work, including him declining to sing one of his songs, commenting that he had mainly translated drinking songs that were probably not suitable for such an occasion.

17 These findings are reported in Ulf Büntgen et al., 'Multi-proxy Dating of Iceland's Major Pre-settlement Katla Eruption of 822–823 CE', *Geology*, 45 (2017), pp. 783–6. This built on earlier work: Kate T. Smith and Hreinn Haraldsson, 'A Late Holocene Jökulhlaup, Markarfljót, Iceland: Nature and Impacts', *Jökull*, 22 (2005), pp. 75–86; and Kate T. Smith and Andrew J. Dugmore, 'Jökulhlaups circa Landnám: Mid- to Late First Millennium AD Floods in South Iceland and Their Implications for Landscapes of Settlement', *Geografiska Annaler*, 88 (2006), pp. 165–76.

18 Clive Oppenheimer et al., 'The Eldgjá Eruption: Timing, Long-range Impacts and Influence on the Christianisation of Iceland', *Climatic Change*, 147 (2018), pp. 369–81.

19 Hervé Guillou et al., '40Ar/39Ar Dating of Thorsmork Ignimbrite and Icelandic Sub-glacial Rhyolites', *Quaternary Science Reviews*, 209 (2019), pp. 52–62.

20 Jules Verne, *Voyage au centre de la terre* (1864), first published in English in 1871 in London by Ward, Lock & Co. Ltd, as *A Journey to the Centre of the Earth*.

21 *Katla* means 'kettle', due to the depression that forms as a result of geothermal melting above the crater, but the volcano is usually referred to as the Witch. The folktale about Katla concerns a housekeeper called Katla, who worked for the abbot of Þykkvabæjarklaustur monastery (at Álftaver on the plain of Mýrdalssandur) and who owned magic trousers. One night she went out to have dinner with the abbot and instructed Barði the shepherd to bring in all the sheep to the homestead before she got back, ready for milking in the morning. It was an impossible task, as the sheep were spread far and wide and up into the mountains. Barði borrowed Katla's magic trousers, allowing him to run far and fast without tiring, and gathered all the sheep, before replacing the trousers. When Katla returned from her night out she saw the sheep were in and the shepherd was asleep, and knew that the only way he could have accomplished this task was to have used her trousers. In her anger she killed him and hid his body in a barrel of whey. Over the winter she realized that his body was going to be exposed as the whey was used up, and she would be found out, so she took the magic trousers, ran up onto the Mýrdalsjökull ice cap and jumped into a big canyon. Shortly after this there was a great flood. Now,

according to folklore, every time that Katla gets angry she melts so much of the ice that there is a big flood that covers all of Mýrdalssandur and the Klaustur area.

22 This was an eventful time for Iceland. Spanish flu arrived on three separate boats from Denmark, the USA and the UK, eventually leading to 540 deaths, 521 in the first six weeks, although the north, east and much of the south escaped infection by effective quarantining. Not long after, on 1 December, Iceland agreed a treaty with Denmark to gain independence, albeit still with the Danish king as head of state – a sovereign nation in a personal union with Denmark.

23 There is good and well-illustrated coverage of the eruption at https://volcano.si.edu, the Global Volcanic Program of the Smithsonian Institution.

24 Reported in October 2019 at https://cineuropa.org.

25 Johann Heinrich Carl Mattheus Ralph Walther von Knebel (1880–1907), to give him his full name, had previously visited Iceland and published papers in 1905, 1906 and 1907 on Icelandic volcanic geology. His work was published posthumously due to the efforts of Hans Reck as *Island. Eine naturwissenschaftliche Studie* (Stuttgart, 1912).

26 Sigríður Sif Gylfadóttir et al., 'The 2014 Lake Askja Rockslide-induced Tsunami: Optimization of Numerical Tsunami Model Using

Observed Data', *JGR Oceans*, 122 (2017), pp. 4110–22.

27 W. G. Lock, *Askja: Iceland's Largest Volcano. With a Description of the Great Lava Desert in the Interior; and a Chapter on the Genesis of the Island* (London, 1881).

28 W.S.C. Russell, 'Askja, a Volcano in the Interior of Iceland', *Geographical Review*, 3 (1917), pp. 212–21. Waterman Spaulding Chapman Russell (1871–1918) was an American who travelled quite widely in Iceland and wrote *Iceland: Horseback Tours in Saga Land*, which was published in Boston in 1914. He dedicated the book to his wife with the words, 'To my wife Grace who twice previously accompanied me over Icelandic trails and twice displayed the greater courage remaining at home alone.' This was accompanied by a frontispiece, a photograph of him and Grace, respectively in Faroese and Icelandic national costume. He did describe Henderson's book as a classic, so his biblical reference is probably to be expected.

29 See www.explorationmuseum.com.

30 See https://noetic.org.

31 Details can be found at NASA, https://nasa.org, as a Research Report and also within details of the Ames Research Center.

32 There are many sources of information on the Heimæy eruption, both with images and video clips. The most detailed overview is a USGS (United States Geological

Survey) booklet by Richard J. Williams Jr and James G. Moore: *Man against Volcano: The Eruption on Heimæy, Vestmannæyjar, Iceland*, originally produced in 1976 and revised in 1983.

33 Accounts of elements of these raids appear in a range of sources. Most recently there has been a book on the subject: Þorsteinn Helgason, *The Corsairs' Longest Voyage: The Turkish Raid in Iceland, 1627* (Leiden, 2018). A novel by the actress Steinunn Johannesdóttir, *Reisubók Guðríðar Símonardóttur* (The Travels of Guðríður Símonardóttir), published in Icelandic in 2002, has yet to be translated into English, although it is available in French, German and Norwegian.

34 Details on NAME can be found on the UK Meteorological website: https://metoffice.gov.uk.

35 The Iceland Volcanoes Supersite can be found at www.geo-nsl.org.

36 This research had quite widespread media coverage, such as in *National Geographic*, with the headline 'Magma Speed Record Set by Icelandic Volcano', and was based on Euan J. Mutch et al., 'Rapid Transcrustal Magma Movement under Iceland', *Nature Geoscience*, 12 (2019), pp. 569–74.

37 Speeds are often quoted in metres per second, or ms^{-1}; 10 ms^{-1} converts to 36 kph (22.4 mph).

38 Data from Páll Einarsson, 'Short-term Seismic Precursors to Icelandic Eruptions', *Frontiers in Earth Science*, 6 (2018), article 45.

39 *Jökulhlaup* is now an accepted term for a glacier flood, although more widely the term Glacial Lake Outburst Flood (GLOF) tends to be used, in which case *jökulhlaup* would only be used for volcanically triggered floods; however, in Iceland a flood from an ice-dammed lake without any volcanic input would probably also be termed *jökulhlaup*. When used as a universal geological term the plural would appear as *jökulhlaups*, but as the plural spelling is the same as the singular in Icelandic, the Icelandic form is used in this book.

40 TV interview reported in *Reykjavík Grapevine*, December 2019 (https://grapevine.is). An overview of volcanic hazards in Iceland is offered by Magnús Tumi Guðmundsson et al., 'Volcanic Hazards in Iceland', *Jökull*, 28 (2008), pp. 251–68.

41 Guðrún Larsen and Jón Eiríksson, eds, *Holocene Tephrochronology Applications in Southern Iceland: Field Guide and Roadlog* (London, 2013). The data from Lögurinn are from Esther Ruth Guðmundsdóttir et al., 'A New High-resolution Holocene Tephrostratigraphy in Eastern Iceland: Improving the Icelandic and North Atlantic Tephrochronology', *Quaternary Science Reviews*, 150 (2016), pp. 234–49. Data from Haukadalsvatn are from David J. Harning et al., 'Marker Tephra in Haukadalsvatn Lake Sediment:

A Key to the Holocene Tephra
Stratigraphy of Eastern Iceland',
Quaternary Science Reviews, 219
(2019), pp. 154–70.

42 David J. Harning et al., 'Holocene
Tephra from Vestfirðir, Iceland',
Quaternary Geochronology, 46
(2018), pp. 59–76.

CHAPTER 3

1 Rev. Jón Steingrímsson, *Fires of the
Earth: The Laki Eruption, 1783–1784*
(Reykjavík, 1998). This translation
of the original Icelandic text, itself
only published in 1908, has a brief
introduction by Guðmundur
E. Sigvaldsson.

2 There is a popular science account
of the eruption and its impacts:
Jeff Kanipe and Alexandra Witze,
*Island on Fire: The Extraordinary
Story of Laki, the Volcano that
Turned Eighteenth Century Europe
Dark* (New York, 2014). A detailed
account of the volcanology can be
found in Thorvaldur Thordarson
and Stephen Self, 'Atmospheric
and Environmental Effects of the
1783–1784 Laki Eruption: A Review
and Assessment', *Journal of Geophysical Research Atmospheres*, 108 (2003).
There are a number of specific papers on the impacts in and beyond
Iceland, with an overview in G. R.
Demarée and A.E.J. Ogilvie, 'Bons
Baiser d'Islande: Climatic, Environmental and Human Dimensions
Impacts of the Lakagigar Eruption
(1783–1784) in Iceland', in *History
and Climate: Memories of the Future*,
ed. Phil D. Jones et al. (New York,
2001), pp. 219–46.

3 Jón Steingrímsson's autobiography could have been lost as it was
given by his family to a relative,
Bishop Steingrímur Jónsson in

Reykjavík, with instructions to
destroy it. He preserved it and
it was eventually published in
Icelandic between 1912 and 1916 as
Ævisaga séra Jóns Steingrímssonar.
The English translation appeared
in 2002 as *A Very Present Help in
Trouble: The Autobiography of the
Fire-priest* (New York, 2002). A
commentary on the autobiogra-
phy can be found in *A History of
Icelandic Literature*, ed. Daisy L.
Neijmann (Lincoln, NE, 2006).

4 The Icelandic terms are *rofabarð*
(sing.) and *rofabörð* (pl.), but
when used in the literature the
anglicizations rofabard, rofabord
or rofbard have been used, with
the addition of an 's' for the plural.
Ideally the original Icelandic term
should be used, as with other
Icelandic terms that have been
adopted in the scientific literature.

5 Sveinn Pálsson, *Ferðabók Sveins
Pálssonar, Dagbækur og Ritgerðir
1791–1797*. Originally published in
Reykjavík in 1945 with a second
edition in 1983, but no English
translation of this text is available.

6 As yet there appears to be no published account in English of the
observations by Jón Sæmundsson.

7 Benjamin Franklin, 'Meteorological Imaginations and Conjectures', *Memoirs of the Literary and
Philosophical Society of Manchester*,

2 (Manchester, 1784), pp. 373–7. A good synopsis of this paper is Richard J. Payne, 'The Meteorological Imaginations and Conjectures of Benjamin Franklin', *North West Geographer*, 10 (2010), pp. 1–7. Franklin's views on meteors were advanced in that he saw them as celestial bodies burning up as they approached the Earth. The year 1783 had seen considerable meteor activity, with five sighted in Britain between 18 August and 29 October. The first of these was described in a letter to the *London Chronicle* as 'an extraordinary meteor . . . whose lustre almost equalled the sun' and was also termed the Great Meteor.

8 Thorvaldur Thordarson, 'The 1783–1785 AD Laki-Grímsvötn Eruptions I: A Critical Look at the Contemporary Chronicles', *Jökull*, 53 (2003), pp. 1–10.

9 Alwyn Scarth, *Vulcan's Fury: Man against the Volcano* (New Haven, CT, 1999).

10 Brian Zambri et al., 'Modeling the 1783–1784 Laki Eruption in Iceland: 2. Climate Impacts', *JGR Atmospheres*, 124 (2019), pp. 6770–90.

11 Anja Schmidt et al., 'Excess Mortality in Europe following a Future Laki-style Icelandic Eruption', *Proceedings of the National Academy of Sciences*, 108 (2011), pp. 15710–15.

12 Translation by Eyþór Halldórsson from Einar Laxness, 'Á 200 ára afmæli Skaftárelda', *Dýnskógar*, 3 (1985), pp. 97–118.

13 See Pálsson, *Ferðabók Sveins Pálssonar*, and (Chief Justice) M. Stephensen, *Kort Beskrivelse over den Nye Vulcans Ildsprudning i Vester-Skaptafjelds-Syssel paa Island i Aaret 1783* (Copenhagen, 1785) (A Short Description of the New Volcanic Eruption in Western Skaftafell District in the Year 1783) which Henderson would have been able to read in the original Danish, although there was an English account in W. J. Hooker's description of his 1809 visit.

14 Ebenezer Henderson's daughter wrote a biography of her father based on his written works and what few papers remained after he encountered an 'inundation' in St Petersburg: Tulia S. Henderson, *Memoir of the Rev. E. Henderson, DD, PhD Including His Labours in Denmark, Iceland and Russia etc., etc.* (London, 1859).

15 William Shakespeare, *As You Like It*, Act II, Scene i, Duke Senior.

CHAPTER 4

1 The 1805 version of the travels by Eggert Ólafsson and Bjarni Pálsson is an abridged and rewritten version of the original and according to the opening sentence rather oddly has them arriving in Iceland in 1800. Their names are published as Messrs Olafsen & Povelsen, and it is this terminology that is usually followed by later non-Icelandic references.

2 William Lord Watts, *Across the Vatna Jökull; or, Scenes in Iceland: Being a Description of Hitherto Unknown Regions* (London, 1876).

3 For an overview, see Sveinn P. Jakobsson and Magnús Tumi Gudmundsson, 'Subglacial and Intraglacial Volcanic Formations in Iceland', *Jökull*, 58 (2008), pp. 179–96.

4 In Laurence Millman's *Last Places: A Journey in the North* (Boston, MA, 2000) the author claims the first ascent was in 1881 by an American, W. L. Howard, who climbed it by means of an anchor and kite, throwing the anchor upwards then climbing on the attached rope with the assistance of a kite. He does not acknowledge his source, although he congratulates Howard for his 'Yankee ingenuity', and it seems likely he was misled. See also www.summitpost.org, which confirms the 1908 ascent as the first climb – and recognized by Ferðafélag Akureyrar (The Akureyri Touring Association) and Ferðafélag Íslands (The Icelandic Touring Association) by a centenary climb on the same date in 2008.

5 An overview of this World Heritage Site, which is a strongly protected nature reserve, can be found at www.ust.is. Sigurður Þórarinsson produced a relatively short introduction to the eruption in *Surtsey: The New Island in the North Atlantic* (New York, 1967), and to recognize the 50th anniversary of the eruption, an update of the geological findings was published as C. Ian Schipper et al., 'The Surtsey Magma Series', *Scientific Reports* (www.nature.com), 5 (2015).

6 NOAA (National Oceanic and Atmospheric Administration, USA) has a good introduction to this topic at https://oceanservice.noaa.gov.

7 Eric P. Achterberg et al., 'Natural Iron Fertilisation by the Eyjafjallajökull Volcanic Eruption', *Geophysical Research Letters*, 40 (2013), pp. 921–6.

8 Joseph M. Licciardi, Mark Kurtz and Joshua M. Curtice, 'Glacial and Volcanic History of Icelandic Table Mountains from Cosmogenic ^{3}He Exposure Ages', *Quaternary Science Reviews*, 26 (2007), pp. 1529–46.

9 T. Thordarson, 'The Widespread *c.* 10ka Saksunarvatn Tephra Is Not a Product of a Single Eruption', Fall Meeting 2014, *American Geophysical Union Abstracts*, ID V24B–04.

10 Helgi Pjetursson, 'The Glacial Palagonite-formation of Iceland', *Scottish Geographical Magazine*, 16 (1900), pp. 265–93.

11 Discussions of tuyas can be found in a range of texts and most recently research has been summarized in the following: J. K. Russell at al., 'Tuya: A Descriptive Genetic Classification', *Quaternary Science Reviews*, 87 (2014), pp. 70–81.

12 On 22 June 2018, the warning was issued jointly by Iceland's Civil

REFERENCES

Protection Agency (Almannavarnir), https://almannavarnir.is, and the Iceland Meteorological Office (Veðurstofa Íslands), https://

en.vedur.is, with a report also in *Iceland Magazine* on the same day giving details of the scale of the problem (https://icelandmag.is).

CHAPTER 5

1 William Morris, Tuesday 26 July 1871, on reaching Geysir.

2 Sketch by Sabine Baring-Gould, July 1862, in his *Iceland: Its Scenes and Sagas* (London, 1863).

3 Sir Cuthbert Edgar Peek (1855–1901) travelled with John M. Coles and Edward Delmar Morgan on two trips visiting the lesser known parts of Iceland, especially the central desert area. These are recounted in John M. Coles (and Edward Delmar Morgan), *Summer Travelling in Iceland: Being the Narrative of Two Journeys across the Island by Unfrequented Routes*, published in 1882 by John Murray (London). Coles was the Map Curator at the Royal Geographical Society (RGS) and Peek presented a paper on crossing Sprengisandur at the Society in 1882. He was a keen astronomer and meteorologist and produced well-respected maps from his travels, but died aged 46 of 'congestion of the brain'. He would have been another good companion for Lord Dufferin as he presented a trophy for shooting to his former university, Cambridge. He is remembered in the annual Cuthbert Peek Award of the RGS, which is awarded for contributions to mapping.

4 J. F. West, *The Journals of the Stanley Expedition to the Faroe Islands and Iceland in 1789* (Tórshavn, 1970–76). Sir John Thomas Stanley, 6th Baronet of Alderley, was a landowner and amateur scientist who visited Iceland with James Wright, Isaac Berners and John Baine. They also climbed Mount Hekla. His journals were translated into Icelandic in 1979. See also J. Wawn, 'John Thomas Stanley and Iceland: The Sense and Sensibility of an Eighteenth-century Explorer', *Scandinavian Studies*, 53 (1981), pp. 52–76.

5 Various papers on Geysir refer to the earlier records and an interesting source available online is Axel Björnsson, 'Development of Thought on the Nature of Geothermal Fields in Iceland', *Proceedings of the World Geothermal Congress* (2005).

6 Professor Carl Vogt (1817–1895) was a scholar of some renown with principal interests in palaeontology and botany. He is noted for an important memoir on *Archaeopteryx*, and as well as being a supporter of Agassiz, his research purported to support Darwin. He put forward a polygenist approach arguing for different races developing along different evolutionary lines, thus the 'White' race was

MOST UNIMAGINABLY STRANGE

separate from the 'Negro', originating from a different 'Ape'.

7 Originally published in 1833, this was republished in 1989 as A. V. Seaton, ed., *The Travel Journals of George Clayton Atkinson in Iceland, Faroes and Westmanna* (Newcastle upon Tyne).

8 Robert Wilhelm Eberhard Bunsen visited Iceland on an expedition led by the traveller and scientist Wolfgang Sartorius von Waltershausen (of palagonite fame), along with 27 horses and five guides. They were unfortunate in arriving at the same time as an outbreak of measles but nevertheless Bunsen achieved much, despite not being seen as a tough enough explorer by the leader, and produced a number of papers on his findings over the next few years. A sense of his experiences can be found in the letters he sent about his trip to his mother, to whom he was devoted: Ralph E. Oester and Karl Freudenberg, 'Bunsen's Trip to Iceland as Recounted in Letters to His Mother', *Journal of Chemical Education*, 18 (1981), pp. 253–60.

9 Alfred Descloiseaux (1817–1897) tends to get overlooked as a companion of the more famous Bunsen but he was a successful mineralogist in his own right, naming a new crystal he found in Iceland Christianite after the Danish king Christian VIII. The mineral had already been named Phillipsite twenty years earlier but the term 'Christianite' is still used occasionally today. Descloiseaux was also possibly the first person to take photographs in Iceland, producing daguerreotypes on his first visit in 1845, when sent to look for minerals by his professor. The daguerreotypes were mainly of Reykjavík but also included ones of basalt cliffs in the northwest from the boat *La Prévoyante*, which was there in association with the French fishing presence in the area. More details of this are in Æsa Sigurjónsdóttir, 'French Photography in Nineteenth-century Iceland', *History of Photography*, 23 (1999), pp. 10–17.

10 Shaul Hurwitz and Michael Manga, 'The Fascinating and Complex Dynamics of Geyser Eruptions', *Annual Review of Earth and Planetary Sciences*, 45 (2017), pp. 31–59.

11 Details of the Society and *Geyser Gazer Sput* can be found at www.geyserstudy.org.

12 Brian Jones et al., 'The Geological History of Geysir, Iceland: A Tephrochronological Approach to the Dating of Sinter', *Journal of the Geological Society*, 164 (2007), pp. 1241–52.

13 Mrs Ethel Alec Tweedie (1862–1940) was born Ethel Brilliana Harley, the daughter of a well-known physician, George Harley FRS, and travelled to Iceland with a female companion, Miss T., and three men, including her brother Vaughan and A.L.T., her future husband Alec Tweedie. When

her husband's marine insurance broking business failed and he died in 1896, shortly followed by the death of her father, she was left destitute with two young sons to raise. She turned to her writing, painting and photography and eventually made a success of her travelling career. She was also an enthusiastic supporter of women's suffrage. Unfortunately, success did not mean that she was finished with tragedy, as she lost one son in the First World War and the second in an RAF flying accident in 1926. In many ways she was the British equivalent of Ida Pfeiffer, a successful independent woman making her way in an environment dominated by men.

14 Steinunn Kristjánsdóttir, 'The Poisoned Arrows of Amor: Cases of Syphilis from 16th-century Iceland', *Scandinavian Journal of History*, 36 (2011), pp. 406–18.

15 Natascha Mehler, 'The Sulphur Trade of Iceland from the Viking Period to the End of the Hanseatic Period', in *Nordic Middle Ages*, ed. Irene Baug, Janicke Larsen and Sigrid Samset Mygland (Bergen, 2015), pp. 193–212. See also notes on the Krýsuvík mine at www.aditnow.co.uk.

16 Grímur Jónsson Thorkelín was a well-known Icelandic-Danish scholar who spent most of his life in Denmark and is best known for his transcriptions of *Beowulf*.

17 The Iceland Deep Drilling Project has a website at https://iddp.is. A more scientific account is Guðmundur Ó. Friðleifsson et al., 'The Iceland Deep Drilling Project at Reykjanes: Drilling into the Root Zone of a Black Smoker Analog', *Journal of Volcanology and Geothermal Research*, 391 (2020), p. 106435.

CHAPTER 6

1 Ebenezer Henderson, 1814, on viewing the eastern margins of Breiðamerkurjökull, in Ebenezer Henderson, *Iceland; or, The Journal of a Residence in That Land, during the Years 1814 and 1815. Containing Observations of the Natural Phenomena, History, Literature, and Antiquities of the Island; and the Religion, Character, Manners and Customs of Its Inhabitants* (Edinburgh and London, 1818), online at www.hathitrust.org.

2 A range of figures can be found but most tend to agree as far as

the broad numbers are concerned. Helgi Björnsson, *The Glaciers of Iceland: A Historical, Cultural and Scientific Overview* (Paris, 2017), has now become essential reading on this topic. The book is expensive and so not easily accessed but is by far the most authoritative source for anything to do with the glaciation of Iceland. A short earlier review is Helgi Björnsson and Finnur Pálsson, 'Icelandic Glaciers', *Jökull*, 58 (2008), pp. 365–86.

3 The biblical quotes used by Henderson presumably come

from the Bible used by the British and Foreign Bible Society, the basis for their translations into different languages. The Icelandic translation dispensed by Henderson was known as *Grútabiblía* or *Hendersonsbiblía* and produced in 1813. The quotes he used when climbing Snæfellsjökull are Jeremiah 18, 'Will the snows of Lebanon fall from the rock of the field? Or the inundating cold flowing waters be exhausted?', and Psalm 102, 'He remaineth the same, and His years shall have no end.' Henderson's enthusiasm for John Milton meant that he sought out the poet and cleric Jón Þorláksson (1744–1819), who had translated *Paradise Lost* into Icelandic. He found the seventy-year-old working in the fields at his farm at Bægisá, in Hörgárdalur, cutting hay, and Jón Þorláksson was delighted at last to meet one of Milton's countrymen. Out of regard for Þorláksson's work, Henderson managed to get the Literary Fund, a British literary organization, to provide the poet with a stipend equivalent to five years of his clerical salary.

4 This was a report that was probably widely published on Burton's return. The specific quote here is from the *Hyde and Glossop Weekly News*, 21 August 1887.

5 Frederick H. H. Howell, 'The Öræfa Jökull, and Its First Ascent', *Proceedings of the Royal Geograph-*

ical Society and Monthly Record of Geography, 14 (1892), pp. 841–50.

6 Photographs are available to view online as part of the Fiske Collection at Cornell University, and are also reproduced in Frank Ponzi, *Howell's Iceland (1890–1901)* (Reykjavík, 2004).

7 More details of the maps can be found at https://islandkort.is.

8 Helgi Björnsson and Finnur Pálsson, 'Radio-echo Soundings on Icelandic Temperate Glaciers: History of Techniques and Findings', *Annals of Glaciology*, 1–10 (2020).

9 A good introduction to the history of glacial theory can be found in Peter G. Knight, *Glacier: Nature and Culture* (London, 2019).

10 Þórður Þorkelsson Vídalín's work was translated into Icelandic only in 1965.

11 Ólafur Ingólfsson, '"On Glaciers in General and Particular" . . . The Life and Works of an Icelandic Pioneer in Glacial Research', *Boreas*, 20 (1991), pp. 79–84. Eventually his work was published in English in 2004 by the Icelandic Literary Society as *Draft of a Physical, Geographical, and Historical Description of Icelandic Ice Mountains on the Basis of a Journey to the Most Prominent of Them in 1792–1794, with Four Maps and Eight Perspective Drawings*, ed. and trans. Richard S. Williams, Jr, and Oddur Sigurðsson.

12 Hans Wilhelmsson Ahlmann, *Land of Ice and Fire* (London, 1938).

13 C. Vincent et al., 'Why Do the Dark and Light Ogives of Forbes

Bands Have Similar Surface Mass Balances?', *Journal of Glaciology*, 64 (2018), pp. 236–46.

14 Sveinn Pálsson, *Ascent of Öræfajökull*, ed. Richard S. Williams and Oddur Sigurðursson, *Jökull*, 46 (1998), pp. 29–33. One of his colleagues suffered from the altitude and went to sleep, being picked up by the others on their descent. The rope was 14.6 m (48 ft) in length, as a fathom in Iceland was equivalent in length to the fathom elsewhere, as for instance in the British Isles.

15 Jónas Hallgrímsson (1807–1845), who was considered the foremost Icelandic Romantic poet, had a love of nature and undertook travels around the country between 1837 and 1845, making geological observations.

16 Bjarni Thorarensen (1786–1841) is described as the first Icelandic Romantic poet and was an influence on Jónas Hallgrímsson. He was brought up at Hlíðarendi, the home of Gunnar in *Njáls saga* (or *Brennu-Njáls saga*).

CHAPTER 7

1 Comment from the Captain of the Icelandic coastguard cutter who took Hans Wilhelmsson Ahlmann and his party from Reykjavík to Hornafjörður in April 1936, quoted by Ahlmann in *Land of Ice and Fire* (London, 1938).

2 Dr Th. Thoroddsen, 'Explorations in Iceland during the Years 1881–98', *Geographical Journal*, 13 (1899), pp. 480–513. In this he footnotes all his previous work and maps published in various Scandinavian journals, in particular some translations into German, which in turn were then translated into English and appeared in earlier reports for the Royal Geographical Society (RGS). In the earlier accounts he made clear the significant challenges he faced working in Iceland, especially in central Iceland. The 1899 paper is illustrated by a number of photographs, including one of

a very evocative lone tent – 'Dr Thoroddsen's Tent'.

3 Although internationally recognized for the quality and breadth of his academic work, as published in a wide range of papers, it was Hans Wilhelmsson Ahlmann's *Land of Ice and Fire*, that brought him to a wider anglophone audience. This was first published in English in 1938, although it originally appeared in 1936 in Swedish as *om Islandsresan* following his exploration of Vatnajökull. In 1939 he was awarded the Patron's Gold Medal of the RGS, one of the two most prestigious awards of the Society, which are given for 'the encouragement and promotion of geographical science and discovery'.

4 Mott T. Greene, *Alfred Wegener: Science, Exploration and the Theory of Continental Drift* (Baltimore, MD, 2015).

MOST UNIMAGINABLY STRANGE

5 Gunnar Hoppe and Valter Schytt, *Memorial to Hans W. Ahlmann, 1889–1974*, Geological Society of America. Obituary in the *Geographical Journal*, 40 (1974), pp. 523–4.

6 Review by R.N.R.B. in *Nature*, 141 (1938), p. 896.

7 Ahlmann seemed to have more respect for the British expeditions, the findings from which appeared in the *Geographical Journal* and the *Alpine Journal*, the publication of the Alpine Club of London, than he had for their German equivalents, although Auden did not seem to be convinced.

8 The meticulous observations of Emmy Todtmann in Iceland have not been widely appreciated outside the German-speaking world. Her major work was produced in 1960 as *Gletscherfor-schungen auf Island (Vatnajökull)*. In the *Biographical Dictionary of Women in Science: Pioneering Lives from Ancient Times to the Mid-20th Century*, ed. Marilyn Ogilvie, Joy Harvey and Margaret Rossiter (2000), she only warrants a bare sentence: 'Emmy Todtmann was known for her descriptive geology of Iceland and Spitsbergen' – a very inadequate synopsis of her contributions in a long academic life.

9 Andrea de Pollitzer-Pollenghi, 'Vatnajökull', *Alpine Journal*, 48 (1936), pp. 257–79.

10 Guðfinna Aðalgeirsdóttir et al., 'Icelandic Glaciers: The Recent Mass Balance and Evolution through 20th and 21st Century', *Geophysical Research Abstracts*, 21 (2019); 'Gains at Hofsjökull Ice Cap', Earth Observatory NASA, https://earthobservatorynasa. gov. An analysis of mass balance changes for a widespread range of fourteen Icelandic glaciers for the period 1945–2017 has revealed their variability, with 1994–2014 being a noticeable time of generally negative mass balances: Joaquín M. C. Belart et al., 'Mass Balance of 14 Icelandic Glaciers, 1945–2017: Spatial Variations and Links with Climate', *Frontiers in Earth Science*, 8 (2020), Article 163.

11 See Jack D. Ives, *Skaftafell in Ice-land: A Thousand Years of Change* (Reykjavík, 2007), and Jack D. Ives, 'Equipment Belonging to Ian Harrison and Tony Prosser Lost on Öræfajökull in 1953, Recovered in July 2006', *Jökull*, 56 (2006), pp. 79–84.

12 Hakon Wadell, 'Vatnajökull: Some Studies and Observations from the Greatest Glacial Area in Iceland', *Geografiska Annaler*, 2 (1920), pp. 300–323.

13 Report of the lecture given in Cambridge on 6 March 1952 in 'Some New Aspects of the Grímsvötn Problem', *Journal of Glaciology*, 2 (1953), pp. 267–75.

14 Kirsten Hastrup, 'Icelandic Topog-raphy and the Sense of Identity', in *Nordic Landscapes: Region and Belonging on the Northern Edge of Europe*, ed. Michael Jones and

Kenneth R. Olwig (Minneapolis, MN, 2008), pp. 53–76.

15 Bureau of Aircraft Accident Archives, www.baaa.acro.com.

16 A recent contribution to this topic is D. I. Benn et al., 'A General Theory of Glacier Surges', *Journal of Glaciology*, 65 (2019), pp. 701–16.

17 These observations have proved invaluable over the years and contribute to the World Glacier Monitoring Service. M. Jackson (see Reference 18) provides an example of how one of her interviewees in Höfn introduced himself as a teacher and glacier monitor, reflecting the value and importance he placed on the role.

18 M. Jackson, *The Secret Lives of Glaciers* (Brattleboro, VT, 2019). There is an interesting article on glacier monitoring in Iceland at www.wired.com: 'Humans Tracked Iceland's Glaciers for Ages. Now, Tech Does' (5 June 2018).

19 Oddur Sigurðsson interview with Jackson, *The Secret Lives of Glaciers*.

20 The English translation by Magnus Magnusson was published in 2005 and includes an introduction by the American writer Susan Sontag, completed a few weeks before her death, describing the novel as 'science fiction, fable, philosophical novel, dream novel, visionary novel, literature of fantasy, wisdom lit, spoof, and sexual turn-on'.

21 Produced in 2007, details can be found at www.katiepaterson.org.

CHAPTER 8

1 Helgi Pjetursson, 'The Glacial Palagonite-formation of Iceland', *Scottish Geographical Magazine*, 16 (1900), pp. 265–93.

2 Oddur Sigurðsson, 'Gláma. Að vera eða vera ekki – jökull', *Náttúrufræðingurinn*, 72 (2014), pp. 47–61. This has an English summary.

3 Friedrich Heinrich Ludwig Konrad Keilhack (1858–1944) first visited Iceland in 1883 on an expedition with Carl W. Schmidt, who was studying rhyolites and went on to make a significant contribution to Icelandic geological research. Keilhack was killed in a bombing raid on Berlin in March 1944. For more on this and other German research see Léo Kristjánsson, 'Historical Notes on Earth Science Research in Iceland by German Expeditions, 1819–1970', in *Geschichte und Philosophie der Geophysik*, ed. G. Schröder (2001), pp. 138–54, available online at https://notendur.hi.is.

4 Baron Carl Wilhelm Paijkull (1836–1869) visited Iceland in 1865 and wrote *En sommer på Island*, translated into English as *A Summer in Iceland* in 1866 by the Rev. M. R. Barnard (vicar of Margaretting in Essex and also author of *Sport in Norway, and Where to Find It*, presumably another book in the libraries of Lord Dufferin and Cuthbert Peek). Paijkull was not only a geologist but also an educationalist,

founding the first Swedish Folk High School. On his early death his wife, described in his Wikipedia entry as the 'intellectual and idealistic' Frederique Paijkull, continued his work. See also https://skbl.se/en.

5 Helgi Pjetursson (1872–1949) gained his PhD in Geology in 1905 and continued with his geological research until 1910. From 1900 onwards he began to turn his attention to more philosophical matters and what he termed astrobiology. Despite him turning away from geology, at an event marking his 70th birthday, the Icelandic Natural History Society (Hið íslenska náttúrufræðifélag) recognized the significance of his research on hyaloclastites by terming him the Nestor of Icelandic natural scientists (Nestor being the legendary wise king of Pylos.)

6 The *Journal of Geology* had a section featuring extended reviews of papers published almost contemporaneously elsewhere and Helgi Pjetursson's paper in the *Scottish Geographical Magazine* was chosen for a very detailed and supportive review by T. C. Chamberlin (8, 1900, pp. 280–83) that ended with the words 'it is gratifying to know that the investigation is likely to be continued.' It was unfortunate that Pjetursson's other interests diverted him from his geological research.

7 The main work by Helgi Pjetursson that formed the basis of his

astrobiological ideas was *Nýall* (New Epoch), published in three volumes in 1919–20 and 1922. Dreaming was central to his theory, as a gateway to life in the universe. This was facilitated by people having a dream-giver, or *draumgjafi* – separate from themselves – who travelled across the universe, between the planets: 'the dream is life, not the remembrance of life.'

8 There is a brief reference to this at https://en.ni.is, and *Grettis saga Ásmundarsonar* is a well-known saga. The story formed the basis of a poem by Louis MacNeice, 'Eclogue from Iceland', that was not part of *Letters from Iceland* but was published a year later, in 1938.

9 Findings from the boulder-watching clubs were published in the *Naturalist Magazine*, which started in 1873, and for a brief period (between 1893 and 1897) in *The Glacialist*, the monthly magazine of glacial geology.

10 See Jón Eiríksson, 'Glaciation Events in the Pliocene-Pleistocene Volcanic Succession of Iceland', *Jökull*, 58 (2008), pp. 315–30, and Áslaug Geirsdóttir, 'Pliocene and Pleistocene Glaciations of Iceland: A Brief Overview of the Glacial History', *Developments in Quaternary Science*, 15 (2011), pp. 199–210.

11 10 m ka⁻¹ means 10 m (33 ft) per thousand years, a is used for year thus a⁻¹ means per year.

12 The Weichselian is the Scandinavian term for the last glacial period, called the Würm in continental

Europe and the Devensian in the British Isles. It covers the period from the last interglacial, the Eemian, to the Holocene, from *c.* 115,000 to 11,700 years ago.

13 Hreggviður Norðdahl et al., 'Late Weichselian and Holocene Environmental History of Iceland', *Jökull*, 58 (2008), pp. 343–62.

14 Johann Striberger et al., 'The Sediments of Lake Lögurinn: A Unique Proxy Record of Holocene Glacial Meltwater Variability in Eastern Iceland', *Quaternary Science Reviews*, 38 (2012), pp. 76–88. D. J. Larsen et al., 'Non-linear Holocene Climate Evolution in the North Atlantic: A High-resolution, Multi-proxy Record of Glacier Activity and Environmental Change from Hvítárvatn, Central Iceland', *Quaternary Science Reviews*, 39 (2012), pp. 14–25.

15 Lagarfljótsormur has a webpage at https://lagarfljotsormur.com, although the last listing there is 2014. There is also an extensive Wikipedia page on the subject.

16 Bergrun A. Oladóttir et al., 'Survival of the Mýrdalsjökull Ice Cap through the Holocene Thermal Maximum: Evidence from Sulphur Contents in Katla Tephra Layers (Iceland) from the Last ~8400 Years', *Annals of Glaciology*, 45 (2007), pp. 183–8.

17 Chris Caseldine and Johann Stötter, '"Little Ice Age" Glaciation of Tröllaskagi Peninsula, Northern Iceland: Climatic Implications for Reconstructed Equilibrium Line

Altitudes (ELAS)', *The Holocene*, 3 (1993), pp. 357–66.

18 Extracts from *Í Suðursveit* are from translations by Kirsten Hastrup, see Kirsten Hastrup, 'Icelandic Topography and the Sense of Identity', in *Nordic Landscapes: Region and Belonging on the Northern Edge of Europe*, ed. Michael Jones and Kenneth R. Olwig (Minneapolis, MN, 2008), pp. 53–76. There is no full English translation of the book yet available. A museum was set up in 2006 close to his birthplace at Hali that includes not only examples of his writings but exhibits relating to the region and the environment in which he grew up, just as Breiðamerkurjökull would have been beginning its long, twentieth-century retreat.

19 Jack Ives produced beautiful maps of the changing margins of southern Vatnajökull in his book on Skaftafell that were based largely on Sigurður Þórarinsson's series of papers in *Geografiska Annaler* in the 1930s and 1940s. These detailed the findings of the Swedish-Icelandic Research Expeditions of 1936, 1937 and 1938. More details on this can also be found in Jean M. Grove, *The Little Ice Age* (London, 1988).

20 Árni Magnússon is best remembered as a collector of manuscripts and an archivist with two collections in Reykjavík and Copenhagen named after him. So too was the character Arne

Saknussemm in Jules Verne's *A Journey to the Centre of the Earth.*

21 Oddur Sigurðsson, 'Gláma. Að vera eða vera ekki – jökull'.

22 Anders Schomacker et al., 'Late Holocene and Modern Glacier Changes in the Marginal Zone of Sólheimajökull, South Iceland', *Jökull*, 62 (2012), pp. 111–30.

23 Cosmogenic or exposure age dating as a geochronological tool only dates back to the mid-1980s and uses in particular ^{10}Be, ^{26}Al, ^{36}Cl and ^{3}He – products of a process known as spallation whereby incoming cosmic rays interact with Si and O in quartz to produce the measured nuclides. Because the rate of build-up of these nuclides in the surface of rocks is not simply a function of the length of exposure of the surface but is dependent on a range of factors, including the rates of cosmic ray production and shielding of the rock, there has been an international attempt to standardize calculations: F. M. Phillips et al., 'The CRONUS-Earth Project: A Synthesis', *Quaternary Geochronology*, 31 (2016), pp. 119–54. Cosmogenic dating has also been used on Martian surfaces.

24 Karianne S. Lilleøren et al., 'The Distribution, Thermal Characteristics and Dynamics of Permafrost in Tröllaskagi, Northern Iceland as Inferred from the Distribution of Rock Glaciers and Ice-cored Moraines', *Permafrost and Periglacial Processes*, 24 (2013), pp. 322–35.

25 The debate over the origins of rock glaciers has been long running, and a recent contribution is Ivar Berthling, 'Beyond Confusion: Rock Glaciers as Cryo-conditioned Landforms', *Geomorphology*, 131 (2011), pp. 98–106. In Iceland some of the latest research is Néstor Campos et al., 'Slow Dynamics in Debris-covered and Rock Glaciers', *Geomorphology*, 342 (2019), pp. 61–77. Earlier work includes W. B. Whalley, 'Rock Glaciers – Permafrost Features or Glacial Relics?', *Permafrost Fourth International Conference Proceedings* (1983), pp. 1396–1401.

26 Bernd Etzelmüller et al., 'Icelandic Permafrost Dynamics since the Last Glacial Maximum: Model Results and Geomorphological Implications', *Quaternary Science Reviews*, 233 (2020), p. 106236.

27 Þorsteinn Sæmundsson et al., 'The Triggering Factors of the Móafellshyrna Debris Slide in Northern Iceland: Intense Precipitation, Earthquake Activity and Thawing of Mountain Permafrost', *Science of the Total Environment*, 621 (2018), pp. 1163–75. Justyna Czekirda et al., 'Transient Modelling of Permafrost Distribution in Iceland', Cryospheric Sciences section of *Frontiers in Earth Science*, https://doi.org/10.3389/feart.2019.00130.

28 Costanza Morino et al., 'Molards as an Indicator of Permafrost Degradation and Landslide Processes', *Earth and Planetary Letters*, 516 (2019), pp. 136–47.

work came with a set of ear plugs. Hilmar Oddson, who directed the film about Jón Leifs, considered his national value to be equivalent to that of Grieg to Norway or Sibelius to Finland, with his music 'growing out of a nature both terrifying and magnificent'. More detail can found in Tony Mitchell, 'New Zealand Glimpsed through Iceland: Music, Place and Psychogeography', *Musicology Australia*, 35 (2013), pp. 41–66.

18 Simon Halink, 'The Icelandic Mythscape: Sagas, Landscapes and National Identity', *National Identities*, 16 (2014), pp. 209–23.

CHAPTER 10

1 William Lord Watts in reflective mood while trudging 'dreamily' across Sprengisandur reflecting on how the landscape he was encountering served the 'important and wonderful mission of connecting the present with the most primitive ages of the world'; see Introduction, Reference 44.

2 From the 'Hetty to Nancy' prose section in Auden and MacNeice, travelling from Hveravellir to Arnarvatn. The quote goes on to say that the stones were 'absolutely unpicturesque and absolutely non-utilitarian'. Travelling over the same area, Ebenezer Henderson had cause to return once again to Jeremiah 2: 'a land of deserts and pits', 'a land of drought', and of 'the shadow of death'.

3 As an introduction to the deserts of Iceland there is Olafur Arnalds, F. O. Gisladottir and H. Sigurjonsson, 'Sandy Deserts of Iceland: An Overview', *Journal of Arid Research*, 47 (2001), pp. 359–71. Recent more detailed research is Olafur Arnalds, Pavla Dagsson-Waldhauserova and Haraldur Olafsson, 'The Icelandic Volcanic Aeolian Environment: Processes and Impacts – A Review', *Aeolian Research*, 20 (2016), pp. 176–95. Pavla Dagsson-Waldhauserova, Olafur Arnalds and Haraldur Olafsson, 'Long-term Dust Aerosol Production from Natural Sources in Iceland', *Journal of the Air and Waste Management Association*, 67 (2017), pp. 173–81.

4 Dragana Đorđević et al., 'Can Volcanic Dust Suspended from Surface Soil and Deserts of Iceland Be Transferred to Central Balkan Similarly to African Dust?', *Frontiers in Earth Science*, 7 (July 2019), article 142.

5 Sigurður Þórarinsson, 'The Öræfajökull Eruption of 1362', *Acta Naturalia Islandica*, 2 (1958), pp. 1–100.

6 Andrew Dugmore et al., 'Abandoned Farms, Volcanic Impacts, and Woodland Management: Revisiting Þjórsárdalur, the "Pompeii" of Iceland', *Arctic Anthropology*, 44 (2007), pp. 1–11.

7 The Eldheimar Museum website is www.eldheimar.is.

8 Andrew J. Dugmore et al.,
'Conceptual Models of 1200 Years
of Icelandic Soil Erosion Reconstructed Using Tephrochronology',
Journal of the North Atlantic, 2
(2009), pp. 1–18.

9 Ólafur Arnalds of the Agricultural
University of Iceland (Land-
búnaðarháskóli Íslands) has been
the main Icelandic researcher in
this area. Publications on the topic
include Ólafur Arnalds et al., *Soil
Erosion in Iceland* (Reykjavík, 2001),
and Ólafur Arnalds, 'Soils of Iceland', *Jökull*, 58 (2008), pp. 409–21.
His brother was Assistant Director
of the Soil Conservation Service
of Iceland (Landgræðslan), and
this brother's son is the composer
Ólafur Arnalds, known to British
audiences for the music to the
TV series *Broadchurch*.

10 Website at https://land.is.

11 Website at www.skogur.is. There
is also a guide to the forests of
Iceland online at https://guide-
toiceland.is.

12 Hannes Þórður Pétursson Hafstein (1861–1922), from Hörgárdalur in northern Iceland, was a
popular poet of the Romantic
tradition as well as a politician
and CEO of Íslandsbanki.

13 It was reported in *The Reykjavík
Grapevine* in February 2020 that
Bishop Agnes Sigurðardóttir of
the National Church, the Evangel-
ical Lutheran Church of Iceland
(hin evangelíska lúterska kirkja),
was planning a 'baptism forest'
(https://grapevine.is).

14 Reported in *Iceland Review*, 26 September 2017 (www.icelandreview.
com).

15 Halldór Sverrisson, 'A Review
of the Icelandic Poplar Breeding
Program', *Technical Review, Iceland
Forest Service* (2017).

16 Reported in various newspapers
and magazines in December 2019
and January 2020 at the time of
the major New Year fireworks
display in Reykjavík.

17 This topic is very well covered in
Karl Benediktsson, 'Floral Hazards: The Nootka Lupin in Iceland
and the Complex Politics of Invasive Life', *Geografiska Annaler B*, 97
(2015), pp. 139–54. The debate also
made the *New York Times* in 2018.

18 This research has shown a range
of potential impacts on plant and
insect diversity, as in Jonathan
Willow, Mariana Tamayo and
Magnús Jóhannsson, 'Potential
Impact of Nootka Lupin (*Lupin
nootkatensis*) Invasion on Polli-
nator Communities in Iceland',
Icelandic Agricultural Sciences, 30
(2017), pp. 51–4. Also Vanessa
Vetter et al., 'Invasion of a Legume
Ecosystem Engineer in a Cold
Biome Alters Plant Diversity',
Frontiers in Plant Science, 9 (2018),
p. 715.

19 By coincidence this was written
on 15 March 2020, the day of the
first 2020 golden plover sighting
at Höfn in southern Iceland. The
average date over recent decades
has been 23 March, although it
was almost a week later between

2017 and 2019. For more details see, for example, 'Spring Has Officially Arrived: The Cultural Significance of the Golden Plover to Icelanders', *Iceland Review*, 29 March 2018 (www.icelandreview.com).

20 Borgný Katrínadóttir et al., 'The Effects of Habitat Type and Volcanic Eruptions on the Breeding Demography of Icelandic Whimbrels, *Numenius phaeopus*', PLOS One, 10 (2015), https://doi.org/10.1371/journal.pone.0131395.

21 *Flóra Íslands* by the botanist Stefán Stefánsson (1863–1921) was first published in 1901 and ran to a number of editions, also being translated into four languages. For a more user-friendly field guide, 1970 saw Áskell Löve's *Íslenzk Ferðaflóra* published, translated into English in 1983 as *Icelandic Excursion Flora*, and more recently *A Guide to the Flowering Plants and Ferns of Iceland* by Hörður Kristinsson, originally published in 1987 and now into its third edition (2013).

22 Jón Guðmundsson, 1574, *Um þau grös og urter, sem vaxa í Íslandi og þeirra dygdir og náttúru* (On Grass and Herbs That Grow in Iceland, Their Virtues and Nature).

23 Johann Gerhard König (1728–1783) was a renowned Baltic German botanist who was born in Livonia, in what is now Latvia, and who studied under Carl Linnaeus. Linnaeus named Iceland's purslane *Koenigia islandica*, after his pupil. He was best known for his

botanical work undertaken while working for the East India Company (EIC) and was considered the founder of modern botany in India. He left his papers to Sir Joseph Banks, who believed he had repaid the EIC 'a thousand-fold by the discovery of drugs and dyeing materials for the European market'.

24 Paul Gaimard, *Voyage en Islande et au Groenland, sur la corvette 'La recherche', en 1835–36*. This was the general account and there were individual reports on various topics associated with the publication.

25 Charles C. Babington, 'List of Plants Gathered during a Short Visit to Iceland in 1846', *Annals and Magazine of Natural History*, 20 (1847), pp. 30–34.

26 W. Lauder Lindsay, 'The Flora of Iceland', *Edinburgh New Philosophical Journal*, 4 (1861), pp. 64–101.

27 Paweł Wąsowicz, 'The First Attempt to List the Archaeotypes of Iceland', *Acta Societatis Botanicorum Poloniae*, 87 (2018), p. 3608.

28 Ingvar Svanberg and Sigurður Ægisson, 'Edible Wild Plant Use in the Faroe Islands and Iceland', *Acta Societatis Botanicorum Poloniae*, 81 (2012), pp. 233–8. *Angelica archangelica* does have medicinal properties, although the claim to cure smallpox from which it is said to get its name – after a monk was showed the plant by an angel in a vision – has regrettably not been proved. The medicinal properties of Icelandic plants have recently

been covered by Anna Rósa Róbertsdóttir, *Icelandic Herbs and Their Medicinal Use* (Berkeley, CA, 2016).

29 Ivar Svanberg, 'The Use of Wild Plants in the Faroe Islands, 1590–1990: A Contribution to Scandinavain Ethnobotany', *Svenska Linnésällskapets Årsskrift*, 1998 (1996–7), pp. 81–130.

30 To quote the advert for Wild Crowberry Wine, 'untamed wine from untamed nature . . . hand-picked from the rustling shrubs of the tundra'.

31 Pawel Wasowicz, Ewa Maria Przedpelska-Wasowicz and Hörður Kristinsson, 'Alien Vascular Plants in Iceland: Diversity, Spatial Patterns, Temporal Trends, and the Impact of Climatic Change', *Flora*, 208 (2013), pp. 648–73.

32 Pawel Wasowicz, 'Non-native Species in the Vascular Flora of Highlands and Mountains of Iceland', PEERJ, 4 (2016), p. 1559.

33 Askell Löve and Doris Löve, eds, *North Atlantic Biota and Their History* (Oxford, 1963). There is a very good review of all the arguments in Hafdís Ægisdóttir and Þóra Ellen Þórhallsdóttir, 'Theories on Migration and History of the North Atlantic Flora', *Jökull*, 54 (2004), pp. 1–16.

34 Inger Greve Alsos et al., 'Long-distance Plant Dispersal to North Atlantic Islands: Colonization Routes and Founder Effect', AOB PLANTS, 7 (2015), plvo36.

https://doi.org/10.1093/aobpla/plvo36.

35 Ólafur Eggertsson, 'Origin of the Driftwood on the Coasts of Iceland: A Study', *Jökull*, 43 (1993), pp. 15–32.

36 Campeachy wood is *Haematoxylum campechianum*, logwood, blackwood or a number of other names, and originates from northern Central America, where it was prized for its dyeing qualities. Cork tree is presumably cork oak, *Quercus suber*, that comes from southwest Europe and northwest Africa, although travelling by currents from these areas to Iceland would involve a very long transit.

37 There are a range of ways in which the bean could be used, from being made into an infusion by grinding into a powder or boiled in water that would then be drunk, to wrapping in flour and white linen and attaching to the woman's stomach. The talismanic quality of sea hearts was a very widespread phenomenon across the North Atlantic. In the Hebrides it was known as the Virgin Mary bean, used as 'an Amulet against Witchcraft, or an Evil Eye', as described in detail in Martin Martin's 1703 *Description of the Western Islands of Scotland*. For Iceland there is a summary about *lausnarsteinn* at https://brodgar.co.uk.

38 Peter Scott and James Fisher, *A Thousand Geese* (London, 1953).

REFERENCES

39 Jose A. Alves, 'Very Rapid Long-distance Sea Crossing by a Migratory Bird', *Nature Scientific Reports*, 6 (2016), p. 3815.

40 Details are available on the UNESCO website, https://whc.unesco.org.

41 Very much from personal experience, for while coring in this lake it was a strange experience to be surrounded by a plateau largely devoid of vegetation yet be engulfed in swarms of midges.

42 More details of the museum can be found at www.fuglasafn.is.

43 Thomas H. McGovern et al., 'Coastal Connections, Local Fishing, and Sustainable Egg Harvesting: Patterns of Viking Age Inland Wild Resource Use in Mývatn District, Northern Iceland', *Environmental Archaeology*, 11 (2006), pp. 187–205.

44 Douglas Hill Scott, *Sportsman's and Tourist's Handbook to Iceland* (Leith, 1899).

45 Sea ice records from 1600 to the present have been particularly well researched, as in Jean Grove, *The Little Ice Age* (London, 1988), following the work of L. Koch, 'The East Greenland Ice', *Meddelelser om Grønland*, 130 (1945), p. 3, and H. Sigtryggsson, 'An Outline of Sea-ice Conditions in the Vicinity of Iceland', *Jökull*, 22 (1972), pp. 1–11. Especially valuable is the work of Astrid Ogilvie, such as her important review of the documentary evidence in A.E.J. Ogilvie, 'The Past Climate and Sea-ice

Record for Iceland, Part 1: Data to AD 1780', *Climatic Change*, 6 (1984), pp. 131–52, and also A.E.J. Ogilvie and Ingibjörg Jónsdóttir, 'Sea Ice, Climate and Icelandic Fisheries in the Eighteenth and Nineteenth Centuries', *Arctic*, 53 (2000), pp. 383–94.

46 See Reference 54.

47 The quote from Loti is '*Il ne revint jamais. Une nuit d'août, là bas, au large de la sombre Islande, au milieu d'un grand bruit de fureur, avaient été célébrées ses noces avec la mer.*' The 2001 translation of the novel by M. Jules Cambon has: 'But he never returned. One August night, off gloomy Iceland, mingled with the furious clamour of the sea, his wedding with the sea was very profound.'

48 Details of the museum are at https://en.visitfjardabyggs.is.

49 A detailed account can be found in Xabier Irujo and Viola G. Miglio, eds, *Basque Whaling in Iceland in the XVII Century* (Santa Barbara, CA, 2015). There is an interesting summary of views of the massacre online at www.snjafjallasetur.is.

50 Sjón (Sigurjón Birgir Sigurðsson), a 'literary maverick' according to a review in *The Guardian*, who wrote Björk's biggest Icelandic hit and was nominated for an Oscar in 2000 for the lyrics to 'I've Seen It All', from the soundtrack of the film *Dancer in the Dark*.

51 Boswell ascribes Samuel Johnson's boast to when he was dining, on 13 April 1780, with (among others)

the Bishop of Chester. Before the meal Johnson was unusually quiet, only venturing to quote the chapter in its entirety. The reason for the brevity of Horrebow's chapters was that his work was a series of rebuttals to an account of Iceland entitled *Nachrichten von Island, Grönland und der Strasse Davis* (News from Iceland, Greenland and the Davis Strait), published posthumously in 1747 by the German Johann Anderson (1674–1746). This had been written in the Blefken tradition: reporting a range of observations, probably made by merchants from Hamburg, without actually having visited the country. There is also a similarly concise chapter on snakes, and insects, in Iceland written by Eric Pontoppidan, the Bishop of Bergen, and published in 1755 in his *Natural History of Norway*, explaining that his cursory coverage of the topic was due to the cold preventing snakes and insects from thriving on the island.

52 Skarphéðinn Þórsson, 'The History of Reindeer in Iceland and Reindeer Study, 1979–1981', *Rangifer*, 4 (1984), pp. 22–38.

53 James Nicol, *An Historical and Descriptive Account of Iceland, Greenland and the Faroe Islands with Illustrations of Their Natural History* (Oxford, 1840). Nicol further stated that 'The poor natives prefer this plant to all other food and gratefully acknowledge that a bountiful Providence sends them bread out of the very stones.'

54 *Cetraria islandica* was an important part of the diet from settlement times, not only as a food supplement but also because of perceived medicinal properties, being used in particular by those suffering from bronchitis, pneumonia or tuberculosis. Adverts for Iceland Moss Cocoa could be found in the frontispieces of books being published in instalments in Britain; one was produced, for instance, with Charles Dickens's *Our Mutual Friend* in 1864–5. Advertising for the schnapps takes advantage of the healthy reputation of *Cetraria islandica*: 'Numerous studies carried out by reputed scientists have shown that the active components in Iceland moss are immunostimulants and can prevent the growth of bacteria and viruses.' (www.reykjavikspirits.is).

CHAPTER 11

1 Part of the inscription on the plaque put in place at the site of the former glacier Ok on 18 August 2019.

2 There is an ever burgeoning literature, both academic and non-academic, on the Anthropocene as both a potential geological epoch and a conceptual tool. A taste of the core geological debate can be found in two papers: Jan Zalasiewicz, Mark Williams and Colin N. Waters, 'Can an Anthropocene Series Be Defined and

Recognized?', *Geological Society of London, Special Publications*, 395 (2014), pp. 39–53 [pro-Anthropocene], and P. L. Gibbard and M.J.C. Walker, 'The Term "Anthropocene" in the Context of Formal Geological Classification', *Geological Society of London, Special Publications*, 395 (2014), pp. 29–37 [anti-Anthropocene]. There are now also journals specializing in the topic, such as *Anthropocene* (started in 2013) and *The Anthropocene Review* (2014).

3 Marcia Bjornerud, *Timefulness: How Thinking Like a Geologist Can Help Save the World* (Princeton, NJ, 2018).

4 Gavin A. Schmidt and Adam Frank, 'The Silurian Hypothesis: Would It Be Possible to Detect an Industrial Civilization in the Geological Record?', *International Journal of Astrobiology*, 18 (2019), pp. 142–50.

5 Richard Streeter et al., 'The Onset of the Palaeoanthropocene in Iceland: Changes in Complex Natural Systems', *The Holocene*, 25 (2015), pp. 1662–75.

6 Monique Gosseling, 'CORDEX Climate Trends for Iceland in the 21st Century', Report, no. VI 2017-009, Veðurstofa Íslands (Icelandic Meteorological Office).

7 A personal example of this from my own research on the climate of the Early Holocene of Iceland: Patrick Moore used the evidence colleagues and myself produced on warmer temperatures during this period as part of his evidence before the U.S. Senate Environment and Public Works Committee to argue that there was no proof that CO_2 caused warming, but that the Earth's climate had been warmer in the recent geological past without greenhouse gas forcing, www.epw.senate.gov. A prime example of a simple climate fact being taken totally out of context.

8 Useful comments on this can be found at www.realclimate.org and at NASA's Earthdata site, https://earthdata.nasa.

9 See Gosseling, 'CORDEX Climate Trends for Iceland in the 21st Century'.

10 Deanne K. Bird and Guðrún Gísladóttir, 'Responding to Volcanic Eruptions in Iceland: From the Small to the Catastrophic', *Palgrave Communications*, 4 (2018), p. 151.

11 Kathleen Compton, Richard A. Bennett and Sigrún Hreinsdóttir, 'Climate-driven Vertical Acceleration of Icelandic Crust Measured by Continuous GPS Geodesy', *Geophysical Research Letters*, 42 (2015), pp. 743–50. This is also summarized in a good piece in *Iceland Review* by Jóhann Páll Ástvaldsson, 'Land Rising Due to Melting Glaciers', 21 August 2019 (www.icelandreview.com).

12 This event got global coverage, with headlines such as 'Scientists Bid Farewell to the First Icelandic Glacier to Be Lost to Climate Change' (CNN) and 'How to Mourn a Glacier' (*New Yorker*).

13 The Swiss glacier gained similar global coverage, as in 'With "Deepest Sadness" Swiss Mourn Passing of Alps Glacier' (*Time*).

14 S. Gosseling, 'CORDEX Climate Trends for Iceland in the 21st Century'.

15 *Heimsljós* was published as a four-part novel between 1937 and 1940; it was translated into English as *World Light* by Magnus Magnusson, published in 2002.

16 My own experience of a Breiðamerkurjökull ice cave was before the current expansion of interest. Travelling to Jökulsárlón with a party of students, the coach driver asked if we would be interested in seeing an ice cave that had just appeared. We stopped to the west of Jökulsárlón and were joined by a 'guide' who led the coach on to a track to the glacier margin and then marched us onto the ice, where he proceeded to dig into a bank of snow, opening up an entrance into a cave. The 'guide' was Hálfdán Björnsson (1927–2017), one of the famous Björnsson family living at Kvísker (especially Flosi Björnsson (1906–1993) and Sigurður Björnsson (1917–2008)), who like Hálfdán were self-taught and nationally known in a range of fields, especially glaciology and natural history. They are referred to in *Öræfi* by Ófeigur Sigurðsson (2014) as the Tvísker brothers.

17 Raynald Harvey Lemelin et al., 'Last Chance Tourism: The Boom, Doom, and Gloom of Visiting Vanishing Destinations', *Current Issues in Tourism*, 13 (2010), pp. 477–93.

18 Details of the earlier avalanches are summarized in 'Avalanches in Iceland' at the Veðurstofa Íslands/ Icelandic Met office website, https://en.vedur.is.

19 'A Large Landslide Falls in Hítardalur Valley', link at https:// en.vedur.is.

20 Herman Farbrot et al., 'Thermal Characteristics and Impact of Climatic Change on Mountain Permafrost in Iceland', *Journal of Geophysical Research: Earth Surface*, 112 (2007), p. F03S90.

21 Steinunn Sigurðardóttir and Heiða Guðný Ásgeirsdóttir, *Heida: A Shepherd at the Edge of the World* (London, 2019), first published in 2016 in Iceland as *Heida: fjalldalabóndinn*.

22 A joint Icelandic-Ukrainian film with an unlikely central character in Halla, Halldóra Gerhardsdóttir, who is both a middle-aged choir director apparently leading a normal urban life and a bike-riding environmental guerilla, 'Mountain Woman', with an ability to bring down power lines with a crossbow and the judicious use of explosives.

23 Haraldur Ólafsson, 'A True Environmental Parable: The Laxá-Mývatn Conflict in Iceland, 1965–1973: An Ecological and Anthropological Approach', *Environmental Review*, 5 (1981), pp. 2–38.

24 First released in 2013; details can be found at www.icelandicfilms. info.

25 When plans were announced in 2015 about possible new hydro-electric power plans, including on Jökulsá á Fjöllum, on which Dettifoss stands, the writer Andri Snær Magnason quoted the 1970 Laxness essay to encapsulate environmental opposition to such developments.

26 Details are at www.carbfix.com, and more of the science can be found in Juerg M. Matter et al., 'The CarbFix Pilot Project: Storing Carbon Dioxide in Basalt', *Energy Procedia*, 2 (2011), pp. 5579–85, and Philip A. E. Pogge von Strandmann et al., 'Rapid CO_2 Mineralisation into Calcite at the CarbFix Storage Site Quantified Using Calcium Isotopes', *Nature Communications*, 10 (2019), p. 1983.

27 To quote the advertising: 'A once in a lifetime experience . . . to float between the two continental plates while enjoying the clarity of the water'. A report in *Iceland Review* (13 March 2017) after a sixty-year-old American died, summarizes some of the issues (www.icelandreview.com).

28 Breaching 2 million visitors triggered a concern that tourism was outstripping the country's ability to cope, and the term 'tourism epidemic' appeared in parts of the media. In an interview in October 2018 María Reynisdóttir, a tourism specialist from the Ministry of Industries and Innovation, disagreed but did accept that tourism was growing at a rate that was not sustainable (www.cntraveler.com).

29 Anna Dóra Sæþórsdóttir, C. Michael Hall and Jarkko Saarinen, 'Making Wilderness, Tourism and the History of the Wilderness Idea in Iceland', *Polar Geography*, 34 (2011), pp. 249–73.

30 Report in *Iceland Review*, and Rannveig Ólafsdóttir and Anna Dóra Sæþórsdóttir, 'Public Perception of Wilderness in Iceland', *Land*, 9 (2020), p. 99.

31 Viscount James Bryce (1838–1922), lawyer and Liberal politician who, like Morris and others, was attracted to Iceland by the sagas and visited in 1872. It provided a chapter in his *Memorials of Travel*, published in London in 1923.

32 Elizabeth Jane Oswald, *By Fell and Fjord; or, Scenes and Studies in Iceland* (Edinburgh, 1882), an account based on travels the previous decade; another visitor attracted by the sagas who was not averse to expressing a candid opinion. She admitted she found some of the scenery 'downright ugly'.

33 In interviews, Sigur Rós tend to play down the linkage between their music and the Icelandic landscape, but a group that can produce a 24-hour live stream of ambient music from a drive around the country on one of the longest days of the year in 2016 (a work entitled *Route One*) cannot avoid most listeners seeing the

landscape as fundamental to what they are hearing.

34 N. Dibben, 'Nature and Nation: National Identity and Environmentalism in Icelandic Popular Music Video and Music Documentary', *Ethnomusicology Forum*, 18 (2009), pp. 131–51. This is a developing field in terms of broader Icelandic music, as in Tony Mitchell, 'Music and Landscape in Iceland', in *The Oxford Handbook of Popular Music in Nordic Countries*, ed. Fabian Holt and Antti-Ville Kärjä (New York, 2017), pp. 145–62, and particularly Þorbjörg Daphne Hall et al., eds, *Sounds Icelandic: Essays on Icelandic Music in the 20th and 21st Centuries* (Sheffield, 2019).

POSTSCRIPT

1 *Timefulness: How Thinking Like a Geologist Can Help Save the World* (Princeton, NJ, 2018).
2 Carl Sagan, *The Demon-haunted World: Science as a Candle in the Dark* (New York, 1996).
3 Richard Fidler and Kári Gíslason, *Sagaland* (Sydney, 2017).

ACKNOWLEDGEMENTS

My interest in Iceland owes its origins to Dr Jack Jarvis, with whom I had
an eye-opening few weeks as a Research Assistant to Breiðamerkurjökull
in 1975 and also in 1977. In the following years I have had the good fortune
to spend time in Iceland with a number of excellent colleagues and many
undergraduate students, from whom I feel sure I learnt more than I ever
gave them.

In particular I would like to thank Professor Pete Langdon and
Dr Naomi Holmes for their collaboration and support over a number of
years and a series of colleagues who assisted on field classes, especially
Dr Robin Cullingford, the late and very much missed Dr Richard Jones,
and Professors Klaus Kuhn, Stephan Harrison and Helen Roberts, not
forgetting the unique and thought-provoking experience of having the poet
Alyson Hallett accompanying us for one of the classes.

I was very fortunate in 1979 in taking a small group to study
Gljúfurárkökull in Tröllaskagi and being welcomed by Hjörtur Eldjárn
Þórarinsson, Sigríður Hafstað and their family at Tjörn, and in later years
by Óskar Gunnarsson and Lene Zachariassen at Dæli. The original contact
was facilitated by Tony Escritt who encouraged the study of Gljúfurárjökull
as an ideal subject for surveying glacier change. Tröllaskagi has always
been a favourite location with memorable periods spent with two very
good friends and colleagues: Professor Hans Stötter from the University
of Innsbruck, who with Ushi Stötter-Frisch has provided both academic
and personal support over almost four decades, and especially Dr Thomas
Häberle, Uncle Chocolate Thomas from Zürich, who has provided the
sort of friendship to me and my family that is beyond measure, as well
as sending seemingly endless food parcels from Switzerland, consisting
entirely of chocolate. Professor Brian Whalley has also helped in
encouraging my research in the north, as have Professors Andy Dugmore
and Kevin Edwards. In Iceland I have also been lucky to be supported in a
variety of ways by a number of colleagues, especially Hreggviður Norðdahl,
and Professors Ólafur Ingólfsson and Áslaug Geirsdóttir at the University

of Iceland, Halldór Pétursson at Akureyri, Oddur Sigurðsson, now at Veðurstofa Íslands and Dr Margrét Hallsdóttir. I can only apologize to them for any errors in the book, which are entirely my responsibility. I am also particularly grateful to Oddur Sigurðsson and Dr Thomas Häberle for providing excellent images, and to Dr Matthew Roberts, Professor Emerita Guðrún Larsen, Professor Brigitte Van Vliet-Lanoë and Professor Emeritus Helgi Björnsson for advice on some of the images.

At Reaktion Books I would like to thank Michael Leaman for his encouragement, and Martha Jay and Alex Ciobanu for their invaluable assistance in putting the book together. To those who read early parts of the book, especially Dr Helen Caseldine, Dr Kate Smith and Dr Thomas Häberle, and Sigurður Gylfi Magnússon, my thanks and sorry if I did not sufficiently 'murder my darlings' as they may have suggested. Mostly I would like to thank my family for putting up with my infatuation with Iceland, usually manifested by their being informed primarily of what was happening to the weather and environment there rather than outside the window, which to them for some reason seemed to matter more. So to Anna, Lucy and Sophie and especially to Helen, my apologies and endless thanks for their tolerance over the years.

PHOTO ACKNOWLEDGEMENTS

The author and publishers wish to express their thanks to the below sources of illustrative material and/or permission to reproduce it. Some locations of artworks are also given below, in the interest of brevity:

Photo Stefan Altenburger, commissioned and produced by Artangel: p. 20; © the artist's estate, courtesy Dafila Scott: p. 256; courtesy Sebastian Ballard: p. 15; from Sabine Baring-Gould, *Iceland: Its Scenes and Sagas* (London, 1863): pp. 144, 188; photos Chris Caseldine: pp. 10–11, 12–13, 28–9, 52–3, 56, 68–9, 71, 88–9, 93, 107, 108, 110–11, 132, 140–41, 145, 170, 202–3, 204 (top), 212, 218–19, 221, 224, 236–7, 244, 248–9, 254, 259, 270, 271, 272–3, 278; photo Dr Helen Caseldine: p. 79; from W. G. Collingwood and Jón Stefánsson, *A Pilgrimage to the Saga-Steads of Iceland* (Ulverston, 1899): p. 45; © 2019 Ólafur Elíasson, courtesy the artist, neugerriemschneider, Berlin, and Tanya Bonakdar Gallery, New York/Los Angeles: pp. 22–3, 24–5; European Union, Copernicus Sentinel-2 imagery: pp. 160–61; Fiske Icelandic Collection, Cornell University Library, Ithaca, NY: pp. 51, 151, 174; from Archibald Fullarton, *The Royal Illustrated Atlas of Modern Geography, with an Introductory Notice by Dr N. Shaw, Secretary to the Royal Geographical Society* (London and Edinburgh, 1862): pp. 40–41, 162 (bottom), 204 (bottom); photo Lachlan Gowen/Unsplash: pp. 80–81; courtesy Dr Thomas Häberle: pp. 120–21, 192–3; © Margrét Hallsdóttir, courtesy Icelandic Institute of Natural History, Garðabær: p. 72; from Georg Hartwig, *The Polar and Tropical Worlds: A Description of Man and Nature in the Polar and Equatorial Regions of the Globe* (Springfield, MA, 1872): pp. 54, 148; from Thulia S. Henderson, *Memoir of the Rev. E. Henderson, D.D., Ph.D., Including his Labours in Denmark, Iceland, Russia . . .* (London, 1859): p. 126; Institute of Earth Sciences, University of Iceland, Reykjavík: pp. 98, 222; Det Kongelige Bibliotek, Copenhagen: p. 167; LAM/stock.adobe.com: p. 183; courtesy Carl Mannerfelt and Helgi Björnsson: p. 173; MartinM303/iStock. com: p. 113; © Myndstef – Myndhöfundasjóður Íslands (The Icelandic Visual Art Copyright Association)/photo courtesy Reykjavík Art Museum: p. 47; Nationaal Archief, The Hague: pp. 226–7; National Museum of

MOST UNIMAGINABLY STRANGE

INDEX